PRACTICE
Workbook

TEXAS HSP Math

Grade 4

Visit *The Learning Site!*
www.harcourtschool.com

Printed in the United States of America

ISBN 13: 978-0-15-356837-4

ISBN 10: 0-15-356837-2

2 3 4 5 6 7 8 9 10 073 16 15 14 13 12 11 10 09 08

Contents

UNIT 1: UNDERSTAND WHOLE NUMBERS AND OPERATIONS

UNIT 2: MULTIPLICATION AND DIVISION FACTS

UNIT 3: MULTIPLY BY 1- AND 2-DIGIT NUMBERS

Chapter 7: Understand Multiplication

Chapter 8: Multiply by 1-Digit Numbers

Chapter 9: Multiply by 2-Digit Numbers

UNIT 4: GEOMETRY

Chapter 10: One- and Two-Dimensional Geometric Figures

Chapter 11: Transformations, Symmetry, and Congruence

Chapter 12: Three-Dimensional Geometric Figures

UNIT 5: DIVIDE BY 1-DIGIT DIVISORS

Chapter 13: Understand Division

Chapter 14: Practice Division

UNIT 6: FRACTIONS AND DECIMALS

UNIT 7: TIME, TEMPERATURE, DATA, AND COMBINATIONS

UNIT 8: MEASUREMENT

Chapter 23: Perimeter and Area

Chapter 24: Volume

SPIRAL REVIEW

Name ____________________

Lesson 1.1

Place Value Through Hundred Thousands

Write each number in two other forms

1. 50,000 + 3,000 + 700 + 5

____

____

2. eight hundred thousand, nine hundred thirty-seven

____

____

3. 420,068

____

____

4. 78,641

____

____

Complete.

5. 290,515 = two hundred ninety ______________, five hundred fifteen = ______________ + 90,000 + ______________ + 10 + 5

6. ______________ + 10,000 + 3,000 + 100 + 80 + 9 = 413,1 ______________ = four hundred thirteen thousand, one ______________ eighty-nine

Write the value of the underlined digit in each number.

7. 70<u>5</u>,239 ______________

8. 4<u>1</u>7,208 ______________

9. 914,3<u>2</u>5 ______________

10. 360,04<u>4</u> ______________

Problem Solving and TAKS Prep

11. In 2005, there were 20,556 Bulldogs registered in the American Kennel Club. What are two ways you can represent the number?

____

____

12. In 2005, the Labrador Retriever was the most popular breed in the American Kennel Club with 137,867 registered. Write the number in two other forms.

____

____

13. What is the value of the digit 9 in 390,215?

A 9 hundred

B 9 thousand

C 90 thousand

D 900 thousand

14. In February, eighty-five thousand, six hundred thirteen people went to the Westminster Dog Show. What is the number in standard form?

F 850,630

G 85,630

H 850,613

J 85,613

Name________________________________

Model Millions

Solve.

1. How many hundreds are in 100,000? ____________

2. How many thousands are in 10,000? ____________

3. How many thousands are in 1,000,000? ____________

4. How many hundreds are in 10,000? ____________

5. How many hundreds are in 1,000,000? ____________

6. How many thousands are in 100,000? ____________

Tell whether each number is large enough to be in the millions or more. Write *yes* or *no*.

7. the number of people at a baseball stadium for one game ____________

8. the distance in miles to the nearest star outside our solar system ____________

9. the number of leaves on the trees in a forest ____________

10. the distance in feet across a swimming pool ____________

11. the number of cars people own in the United States ____________

12. the number of trips a bus might make in one day ____________

13. the number of bags of trash a family makes in one month ____________

14. the distance in miles from one city to another in your state ____________

15. the number of students in the United States ____________

16. the number of miles you might travel to reach the Sun ____________

17. the number of gallons of water in the ocean ____________

18. the number of stars in the Milky Way ____________

Choose the number in which the digit 5 has the greater value.

19. 435,767 or 450,767 ____________

20. 510,000 or 5,100,000 ____________

21. 125,000,000 or 521,000,000 ____________

22. 435,003 or 4,300,500 ____________

23. 1,511,672 or 115,672 ____________

24. 40,005,400 or 350,400,300 ____________

25. 135,322,000 or 9,450,322 ____________

26. 35,000,000 or 3,500,000 ____________

Name________________________________ Lesson 1.3

Place Value Through Millions

Write each number in two other forms.

1. ninety-five million, three thousand, sixteen

____

2. four hundred eighty-five million, fifty-two thousand, one hundred eight

____

3. 507,340,015

____

____

____

4. 20,000,000 + 500,000 + 60,000 + 1,000 + 300 + 40

____

____

Use the number 78,024,593.

5. Write the name of the period that has the digits 24. ______________

6. Write the digit in the ten millions place. ______________

7. Write the value of the digit 8.

____

8. Write the name of the period that has the digit 5. ______________

Find the sum. Then write the answer in standard form.

9. 7 thousands 3 hundreds 4 ones + 8 ten thousands 1 thousand 5 hundreds

____

Problem Solving and TAKS Prep

10. The average distance from Earth to the Sun is 92,955,807 miles. What is the value of the digit 2?

____

____

11. The average distance from Earth to the Sun is one hundred forty-nine million, six hundred thousand kilometers. Write the number in standard form.

____

12. What is the value of the digit 8 in 407,380,510?

A 8,000,000 C 80,000
B 800,000 D 8,000

13. What is the value of the digit 4 in 43,902,655?

F 400 thousand H 40 million
G 4 million J 400 million

Name____

Lesson 1.4

Compare Whole Numbers

Use the number line to compare. Write the lesser number.

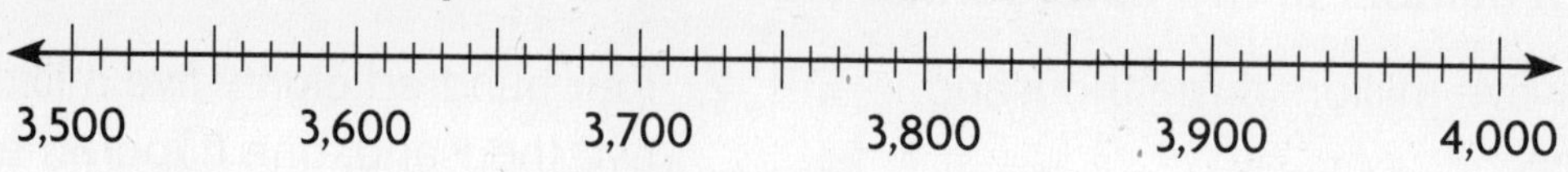

1. 3,660 or 3,590 ____________

2. 3,707 or 3,777 ____________

3. 3,950 or 3,905 ____________

Compare. Write $<$, $>$, or $=$ in each ○.

4. 5,155 ○ 5,751

5. 6,810 ○ 6,279

6. 45,166 ○ 39,867

7. 72,942 ○ 74,288

8. 891,023 ○ 806,321

9. 673,219 ○ 73,551

10. 3,467,284 ○ 481,105

11. 613,500 ○ 1,611,311

12. 4,000,111 ○ 41,011

ALGEBRA Find all of the digits that can replace each □.

13. 781 $\neq$ 78□ ____________

14. 2,4□5 $\neq$ 2,465 ____________

15. □,119 $\neq$ 9,119 ____________

Problem Solving and TAKS Prep

USE DATA For 16–17, use the table.

16. Which mountain is taller, Logan or McKinley?

____

17. Which mountain is taller than 29,000 feet?

____

Tallest Mountains

Mountain	Height (in feet)
Everest	29,028
McKinley	20,320
Logan	19,551

18. Which number is the greatest from the list below?

A 34,544

B 304,544

C 43,450

D 345,144

19. Ernie wants to raise $140 a week for the school fund drive. In four weeks, he's raised $147, $129, $163, and $142. Which total was less than Ernie's weekly goal?

F $147

G $163

H $129

J $142

Name_______________________________

Order Whole Numbers

Write the numbers in order from greatest to least.

1. 74,421; 57,034; 58,925

____

2. 2,917,033; 2,891,022; 2,805,567

____

3. 409,351; 419,531; 417,011

____

4. 25,327,077; 25,998; 2,532,707

____

5. 621,456; 621,045,066; 6,021,456

____

6. 309,423; 305,125; 309,761

____

7. 4,358,190; 4,349,778; 897,455

____

8. 5,090,115; 50,009,115; 509,155

____

ALGEBRA Write all of the digits that can replace each □.

9. $389 < 3\square 7 < 399$

____

10. $5{,}601 < 5{,}\square 01 < 5{,}901$

____

11. $39{,}560 > 3\square{,}570 > 34{,}580$

____

12. $178{,}345 > 1\square 8{,}345 > 148{,}345$

____

Problem Solving and TAKS Prep

USE DATA For 13–14, use the table.

Largest Lakes (area in square miles)	
Victoria	26,828
Huron	23,000
Superior	31,700
Caspian Sea	19,551

13. Which lake has the smallest area?

____

14. Write the names of the lakes in order from least area to the greatest area.

____

15. Automobile sales for four weeks are \$179,384, \$264,635, \$228,775, and \$281,413. Which amount is the greatest?

____

16. Which shows the numbers in order from greatest to least?

A 92,944; 92,299; 92,449

B 159,872; 159,728; 159,287

C 731,422; 731,242; 731,244

D 487,096; 487,609; 487,960

Problem Solving Workshop Strategy: Use Logical Reasoning

Problem Solving Strategy Practice

Use logical reasoning to solve.

1. The stadium store sells team shirts on Friday, Saturday, and Sunday. The number of shirts sold for three days were 473, 618, and 556. The least number of shirts were sold on a Friday. More than 600 shirts were sold on Saturday. How many shirts were sold each day?

____

2. Anton, Rachel, and Lamont like different baseball teams. The teams are the Yankees, the Red Sox, and the White Sox. Anton's favorite team does not have a color in its name. Lamont does not like the White Sox. Which team does each person like best?

____

Mixed Strategy Practice

3. Beth, Paulo, Lee, Maya, and Rob are standing in line to get into the movies. Beth is in front of Maya. Maya is not last in line. Rob is first. Lee is after Maya. Paulo is not last. In what order are they standing in line?

____

4. Mr. Katz bought a signed Alex Rodriguez baseball for $755. He used $50-bills, $20-bills, and $5 bills to make exactly $755. The total number of bills he used is less than 20. What combination of bills would Mr. Katz have used?

____

USE DATA For 5–6, use the information shown in the art.

5. Claire buys two items. She spends less than $100 for both of them. Which two items does she buy?

____

6. Alex wants to save money to buy the hockey stick. He already has $8. He saves twice the amount of money each week. After 2 weeks he has $40. How long do you think it will take Alex to save $72?

____

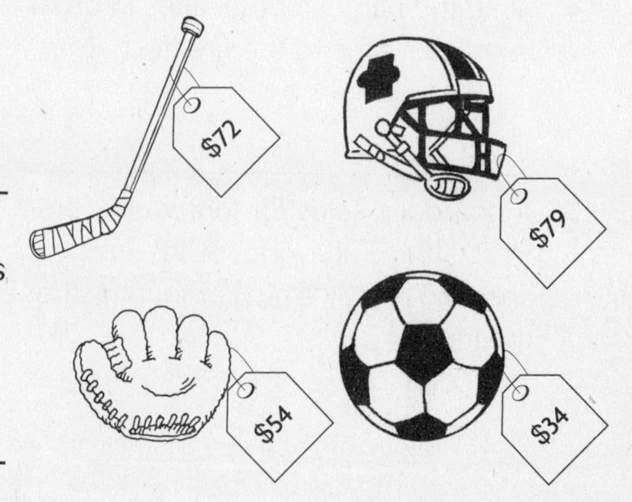

Name ______________________

Algebra: Relate Addition and Subtraction

Write a related fact. Use it to complete the number sentence.

1. $\square - 7 = 8$ ______

2. $4 + \square = 13$ ______

3. $\square + 9 = 14$ ______

4. $8 + \square = 11$ ______

5. $\square - 4 = 8$ ______

6. $17 - \square = 9$ ______

7. $\square - 5 = 5$ ______

8. $13 - \square = 5$ ______

9. $\square + 7 = 16$ ______

Write the fact family for each set of numbers.

10. 6, 8, 14

11. 7, 5, 12

12. 9, 6, 15

Problem Solving and TAKS Prep

13. Byron can do 12 pull-ups. Malik can do 7 pull-ups. How many more pull-ups can Byron do than Malik? What related facts can you use to solve this problem?

14. Byron can do 12 pull-ups. Malik can do 7 pull-ups. Selma does more pull-ups than Malik but fewer than Byron. What are the four possible numbers of pull-ups that Selma could have done?

15. Which of the following sets of numbers cannot be used to make a fact family?

A 25,10,15
B 2,2,4
C 15,9,6
D 7,2,15

16. Which of the following sets of numbers can be used to make a fact family?

F 5,6,11
G 11,12,13
H 7,6,12
J 19,9,11

Name____________________ **Lesson 2.2**

Round Whole Numbers to the Nearest 10, 100, and 1,000

Round each number to the place value of the underlined digit.

1. 7,803 __________
2. 4,097 __________
3. 23,672 __________
4. 627,432 __________
5. 34,809,516 __________
6. 671,523,890 __________

Round each number to the nearest ten, hundred, and thousand.

7. 6,086,341 __________ __________ __________
8. 79,014,878 __________ __________ __________
9. 821,460,934 __________ __________ __________

Problem Solving and TAKS Prep

USE DATA For 10–11, use the table.

10. Which state has a population that rounds to 5,364,000?

____

11. What is the population of Maryland rounded to the nearest thousand?

____

Population of States in 2000 Census

State	Population
Maryland	5,296,486
Tennessee	5,689,283
Wisconsin	5,363,675

12. Jen's brother drove about 45,000 miles last year. Which could be the exact number of miles he drove?

   A 44,399

   B 44,098

   C 44,890

   D 45,987

13. A number rounded to the nearest thousand is 560,000. What is the missing digit? 560,☐95

____

Name ______________________

Lesson 2.3

Mental Math: Addition and Subtraction Patterns

Use mental math to complete the pattern.

1. ______ + 8 = 17
   90 + ______ = 170
   900 + 800 = ______
   9,000 + 8,000 = ______

2. ______ − 4 = 8
   120 − 40 = ______
   1,200 − ______ = 800
   12,000 − 4,000 = ______

3. ______ − 3 = 7
   100 − ______ = 70
   ______ − 300 = 700
   10,000 − 3,000 = ______

4. 7 + 9 = ______
   70 + ______ = 160
   700 + 900 = ______
   ______ + 9,000 = 16,000

5. 8 + ______ = 11
   80 + ______ = 110
   ______ + 300 = 1,100
   ______ + 3,000 = 11,000

6. ______ − 5 = 9
   140 − 50 = ______
   1,400 − ______ = 900
   ______ − 5,000 = 9,000

Use mental math patterns to find the sum or difference.

7. 600 + 700 ______

8. 180 − 90 ______

9. 6,000 + 9,000 ______

10. 13,000 − 5,000 ______

11. 12,000 + 10,000 ______

12. 700 − 600 ______

13. 130,000 + 70,000 ______

14. 15,000 − 8,000 ______

Problem Solving and TAKS Prep

15. In 2001, there are 400 rabbits at the zoo. In 2002, there are 1,200 rabbits at the zoo. How many more rabbits are at the zoo in 2002 than 2001?

______

16. There are 600 pens in each box. How many total pens are there in 2 boxes?

______

17. What number completes the sentence
■ + 3,000 = 12,000?

A 90,000
B 9,000
C 8,000
D 900

18. There are 14,000 newspapers printed on Tuesday morning. By Tuesday afternoon, only 8,000 are sold. How many newspapers have not been sold yet?

______

Name____ Lesson 2.4

Mental Math: Estimate Sums and Differences

Use rounding to estimate.

1. 6,356 + 1,675 = ______

2. 8,267 − 2,761 = ______

3. 38,707 + 28,392 = ______

4. 75,428 − 19,577 = ______

5. 187 + 519 = ______

6. 6,489 − 1,807 = ______

7. 24,655 + 51,683 = ______

8. 61,075 − 29,732 = ______

Use compatible numbers to estimate.

9. 5,432 − 652 ______

10. 45,221 + 6,167 ______

11. 392 + 47 + 89 ______

Adjust the estimate to make it closer to the exact sum or difference.

12. 6,285 + 2,167
Estimate: 8,000 ______

13. 42,819 − 11,786
Estimate: 30,000 ______

14. 17,835 + 45,199
Estimate: 65,000 ______

Problem Solving and TAKS Prep

15. On Saturday, 3,251 people visit the San Diego Zoo. On Sunday, there are 2,987 visitors. About how many people in all visit the zoo on Saturday and Sunday?

______

16. A plane travels 5,742 km this week and 1,623 km next week. About how many more kilometers does the plane travel this week than next week?

______

17. Carolyn buys a car with 13,867 miles on it. In one year, she drives 9,276 miles. Which is the best estimate of how many miles are on the car now?

A 23,000 miles
B 24,000 miles
C 25,000 miles
D 26,000 miles

18. During his daily walk, Eddie normally takes 1,258 steps. Estimate the number of steps Eddie takes on Tuesday if he takes 29 more steps than normal.

______

Mental Math Strategies

Add or subtract mentally. Tell the strategy you used.

1. $73 + 15$ ____________

2. $87 - 48$ ____________

3. $57 + 91$ ____________

4. $152 - 68$ ____________

5. $542 + 148$ ____________

6. $515 - 151$ ____________

7. $799 - 231$ ____________

8. $387 + 73$ ____________

9. $945 - 425$ ____________

10. $452 + 339$ ____________

11. $396 + 265$ ____________

12. $594 - 496$ ____________

Problem Solving and TAKS Prep

13. Vicky has 32 baseball cards and 29 soccer cards. Use mental math to find how many cards Vicky has in all.

____

14. Kareem bowls 78 the first game and 52 the second game. Use mental math to find the difference of Kareem's scores.

____

15. Jason sells 27 tickets on Monday and 34 on Tuesday. He adds 3 to 27 to find the sum mentally. How should he adjust the sum to find the total?

A Add 3 to the sum

B Add 4 to the sum

C Subtract 3 from the sum

D Subtract 4 from the sum

16. Haley buys a baseball bat and glove that cost $25 and $42. She subtracts $2 from $42 to find the total mentally. How should Haley adjust the sum to find the total?

F Add $2 to the sum

G Subtract $2 from the sum

H Add $5 to the sum

J Subtract $5 from the sum

Name________________________________ Lesson 2.6

Problem Solving Workshop Skill: Estimate or Exact Answer?

Problem Solving Skill Practice

Explain whether to estimate or find an exact answer. Then solve the problem.

1. A plane has 5 seating sections that can hold a total of 1,175 passengers. Today, the sections held 187, 210, 194, 115, and 208 passengers. Was the plane filled to capacity?

____

____

2. A small plane carries 130 gallons of fuel. It needs 120 gallons to fly a 45-mile trip. Does the pilot have enough fuel to make a 45-mile trip?

____

____

3. A movie theater has a total of 415 seats. There are 187 adults and 213 children seated in the theater. How many empty seats are there in the theater?

____

____

4. Bob drives 27 miles round trip each day for three days. Has Bob traveled more or less than 250 miles?

____

____

Mixed Applications

5. The movie theater sells 213 tickets on Monday, 187 tickets on Tuesday, and 98 tickets on Wednesday. Are there more, or less than 600 tickets sold for all three days?

____

6. The movie theater sells 209 tickets for "Canyon Trail" and 94 tickets for "A Light in the Sky". How many more tickets are sold at the theater for "Canyon Trail" than "A Light in the Sky"?

____

7. Sara sells 87 tickets for a school benefit. Josh sells 43 tickets. Marc sells 28 tickets. How many more tickets does Sara sell than Marc and Josh together?

____

8. A stamp album contains 126 stamps. Another album contains 67 stamps. Each album can hold up to 150 stamps. How many more stamps can both albums hold altogether?

____

Name______________________________ Lesson 2.7

Add and Subtract Through 4-Digit Numbers

Estimate. Then find the sum or difference.

1. 414 + 727

2. 784 − 149

3. 5,305 + 848

4. 7,322 − 616

5. 2,673 + 4,548

6. 3,357 + 1,219

7. 8,452 − 2,621

8. 9,344 − 5,667

9. 4,955 + 978

10. 9,999 − 901

11. 7,593 + 1,475

12. 8,891 − 1,490

13. 3,069 + 956

14. 6,560 − 5,699

15. 1,948 − 1,052

16. 7,326 + 2,673

ALGEBRA Find the missing digit.

17. 9□8 + 247 = 1,175

18. 7,895 − 1,23□ = 6,661

19. □,689 − 726 = 3,963

20. 1,357 + 7□6 = 2,113

Problem Solving and TAKS Prep

21. Jan drove 324 miles on Monday, then 483 miles on Tuesday. How many miles did Jan drive in all?

____

22. A baseball team scores 759 runs in a season. The next season the team scores 823 runs. How many runs are scored in all?

____

23. An airplane will fly a total of 4,080 miles this trip. The plane has flown 1,576 miles so far. How many more miles will the plane need to travel?

A 2,504 more miles

B 2,514 more miles

C 2,594 more miles

D 5,656 more miles

24. There are 5,873 soccer fans at the first game. There are 3,985 fans at the second game. How many more fans are at the first game? Explain.

____

____

____

Name____

Subtract Across Zeros

Estimate. Then find the difference.

1. 3,078 − 678

2. 760 − 194

3. 6,004 − 452

4. 7,030 − 4,265

5. 8,056 − 2,109

6. 9,000 − 2,708

7. 4,890 − 1,405

8. 6,902 − 3,440

9. 670 − 413 ______

10. 4,700 − 876 ______

11. 5,030 − 2,125 ______

Choose two numbers from the box to make each difference.

4,200	4,000	3,020	3,402	424

12. 3,776 ______

13. 1,180 ______

14. 2,596 ______

15. 598 ______

Problem Solving and TAKS Prep

16. One of the largest volcanic eruptions occurred in 1883 on the Indonesian Island of Krakatoa. How many years before 2006 had this eruption occured?

______

17. Jessie estimates the distance from New York to San Diego to be 3,000 miles. The actual distance is 2,755 miles. What is the difference between Jessie's estimate and the actual distance?

______

18. Helena starts a trip with 4,345 miles on her car. She finishes the trip with 8,050 miles on her car. How many miles did Helena travel on her trip?

A 12,395
B 4,705
C 3,805
D 3,705

19. A mountain peak reaches 3,400 feet in elevation. A mountain climber has climbed 1,987 feet so far. How many more feet does the climber need to go before reaching the top of the peak?

______

Name______________________________

Choose a Method

Find the sum or difference. Write the method you used.

1. 256,684 + 157,925 = ______

2. 845,002 − 32,000 = ______

3. 5,369,021 + 1,488,627 = ______

4. 390,451 − 189,693 = ______

5. 4,244,500 + 110,001 = ______

6. 7,056,634 + 869,378 = ______

7. 5,351,842 − 1,409,876 = ______

8. 6,411,809 − 411,809 = ______

ALGEBRA Find the missing digit.

9. 32□,164 + 651,247 = 974,411

10. 722,□85 − 134,761 = 588,124

11. 314,678 − 1□2,657 = 182,021

12. 7,1□9,236 + 1,292,459 = 8,481,695

Problem Solving and TAKS Prep

13. Jupiter's moon, Callisto, is 1,883,000 kilometers away from Jupiter. Jupiter's other moon, Ganymede, is 1,070,000 kilometers away from Jupiter. What is the difference of these two distances?

____

14. Jessie scores 304,700 points in a video game. Raquel scores 294,750 points. How many more points does Jessie score than Raquel scores?

____

15. Plane A travels 108,495 miles. Plane B travels 97,452 miles. How many miles do both planes travel in all?

   A 195,847 miles
   B 205,847 miles
   C 205,887 miles
   D 205,947 miles

16. Last year, 456,197 fans attended a minor league's baseball games. This year, 387,044 fans attended. What is the total number of fans that attended this year and last year all together?

____

Name______________________________________ **Lesson 3.1**

Addition Properties

Find the missing number. Tell which property you used.

1. $\square + 0 = 0 + 23$ ____________
2. $15 + 5 = \square + 15$ ____________
3. $12 + (2 + 7) = (\square + 2) + 7$ ____________
4. $\square + 7 = 7 + 36$ ____________
5. $\square + 45 = 45 + 0$ ____________
6. $(22 + \square) + 11 = 22 + (44 + 11)$ ____________

Change the order or group the addends so that you can add mentally. Find the sum. Name the property you used.

7. $120 + 37 + 280$ ____________ ____________
8. $25 + 25 + 30$ ____________ ____________
9. $60 + 82 + 40$ ____________ ____________
10. $28 + 21 + 32 + 19$ ____________ ____________
11. $66 + 27 + 44$ ____________ ____________
12. $133 + 25 + 247$ ____________ ____________
13. $45 + 22 + 25$ ____________ ____________
14. $61 + 57 + 39 + 23$ ____________ ____________

Problem Solving and TAKS Prep

USE DATA For 15–16, use the table.

15. Use the Associative Property to find the total number of marbles in Sam's collection.

____

16. Sam buys another 15 Shooter Stripes. How many marbles are in Sam's collection now?

____

Sam's Marble Collection	
Type	**Number**
Blue Chinese Checkers	32
Cat's Eyes	81
Speckled Spots	18
Shooter Stripes	59

17. Which shows the Identity Property of addition?

   A $16 + 0 = 16$
   B $12 + 1 = 13$
   C $29 + 29 = 58$
   D $1 + 1 = 2$

18. Which shows the Commutative Property of addition?

   F $11 + 9 = 20$
   G $0 + 7 = 0$
   H $20 + 20 = 40$
   J $5 + 7 = 7 + 5$

Name ____________________

Write and Evaluate Expressions

Find the value of each expression.

1. $12 - (4 + 3)$ ________

2. $5 + (15 - 3)$ ________

3. $17 - ■$ if $■ = 8$ ________

4. $5 + (m - 2)$ if $m = 12$ ________

5. $(18 + 22) - 15$ ________

6. $(31 - 16) - 8$ ________

7. $■ + 25$ if $■ = 9$ ________

8. $b - (31 + 5)$ if $b = 52$ ________

Write an expression with a variable. Tell what the variable represents.

9. Sally gave away 5 apples.

____

10. Ali had 9 fish and bought some more.

____

11. Theresa put $15 in her bank account.

____

12. Glenn gave away some of his 20 pins.

____

Write words to match each expression.

13. $t - 5$

____

____

____

14. $12 + k$

____

____

____

Problem Solving and TAKS Prep

15. Write words to match the expression $t + 3 - 1$ where t stands for tomatoes for a salad.

____

____

____

____

16. Write words to match the expression $y + 4$ where y stands for the time Josef practiced the piano on Saturday.

____

____

____

____

17. Edie ran 2 miles more than Joan. Which expression shows how far Joan ran?

A $e + 2$

B $2 + e$

C $e + 3$

D $e - 2$

18. There are 6 kittens in a closet. The mother cat removes 3. Write an expression that shows how many kittens are left in the closet.

____

Name______________________________ Lesson 3.3

Addition and Subtraction Equations

Write an equation for each. Choose the variable for the unknown. Tell what the variable represents.

1. Rickie has 15 model cars. Some are red and 8 are blue.

____

2. Wendy had $12. Her mother gave her some more so she now has $17.

____

Solve the equation.

3. $19 - 4 = n$

$n =$ ____

4. $6 + ■ = 19$

$■ =$ ____

5. $r - 12 = 21$

$r =$ ____

6. $t + 14 = 31$

$t =$ ____

Write words to match the equation.

7. $b + 5 = 12$

____

____

8. $a - 9 = 2$

____

____

9. $16 - w = 4$

____

____

10. $y + 7 = 29$

____

____

Problem Solving and TAKS prep

11. Eight hearing dogs graduated in February, 5 in May, and 9 in November. Write and solve an equation that tells how many hearing dogs graduated in all.

____

12. Thirteen dogs graduated in May. There were 5 hearing dogs, 4 service dogs and some tracking dogs. Write an equation that shows the total number of dogs that graduated in May.

____

13. Jed watched 10 minutes of previews and a 50-minute dog movie. Which equation that tells the total time Jed was in the theater.

A $10 + 50 = t$ **C** $t - 10 = 50$

B $50 - t = 10$ **D** $t + 10 = 50$

14. Haley's favorite picture book is 27 pages. 11 of the pages have pictures of dogs. The rest have pictures of birds. Which equation can be used to find how many pages have birds?

F $27 + 11 = b$ **H** $b - 11 = 27$

G $27 - b = 11$ **J** $b + 11 = 27$

Name____ Lesson 3.4

Problem Solving Workshop Strategy: Work Backward

Problem Solving Strategy Practice

Work backward to solve.

1. Leon arrived at the preserve at 11:00 A.M. He began the morning by taking 45 minutes to feed his pets at home and driving 2 hours to get to the preserve. What time did Leon begin?

____

2. Kit read a 25-page book about lions. Seven pages were about hunts, 15 about habitat, and the rest were about prides. How many pages were about prides?

____

3. Twelve lions in the pride did not go on a hunt. When more lions returned from the hunt, there were 21. How many lions were on the hunt?

____

4. Polly ate lunch and then took 15 minutes to walk to Cher's. They rode bikes for 35 minutes and then studied for 20 minutes. If they finished at 2:30, when did Polly finish lunch?

____

Mixed Strategy Practice

5. Five prides were sent from the zoo to a preserve. Two prides were returned. Now there are 17 prides at the zoo. How many prides were there at the zoo before the 5 were sent away?

____

6. Red, blue, green, and brown teams lined up for their assignments. The brown team was ahead of the red team. The blue team was not last. The green team was first. Which team was last?

____

7. **USE DATA** Use the information in the table below to draw a bar graph.

Preserve Lion Population	
Age	**Number**
Cubs	18
Adolescents	14
Mature	2
Older	7

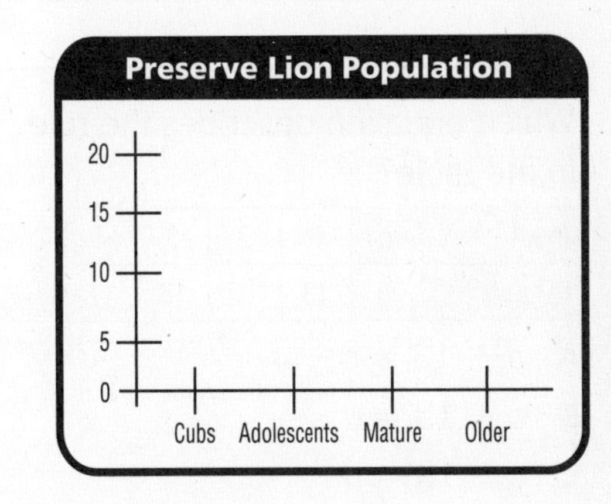

Name____________________ **Lesson 3.5**

Ordered Pairs in a Table

Find a rule. Use your rule to find the next two ordered pairs.

1.

Input	f	10	15	20	25	30
Output	g	5	10	15		

____

2.

Input	c	88	86	84		
Output	d	66	64	62	60	58

____

3.

Input	s			9	5	1
Output	t	70	66	62	58	54

____

4.

Input	x	15	14	13	12	11
Output	y			28	27	26

____

Use the rule to make an input/output table.

5. Add 7 to m.

Input	m				
Output	n				

6. Subtract 14 from a.

Input	a				
Output	b				

Problem Solving and TAKS Prep

USE DATA For 7–8, use the input/output table.

7. A figure is made of a row of squares. One square has a perimeter of 4. Two squares has a perimeter of 6, and so on. Finish the input-output table to show the pattern.

Input	s	1	2	3	4	5
Output	p	4	6			

____

8. What will be the perimeter of 10 squares in a row?

____

9. Which equation describes the rule in the table?

Input	c	0	2	3	4
Output	d	13	15	16	17

A $d + 13 = c$

B $c + 13 = d$

C $c - 13 = d$

D $d - 13 = c$

10. Write a rule for the table.

Input	r	0	1	3	5	7
Output	s	4	5	7	9	11

____

Name____

Lesson 4.1

Algebra: Relate Addition and Multiplication

Write related addition and multiplication sentences for each.

1.

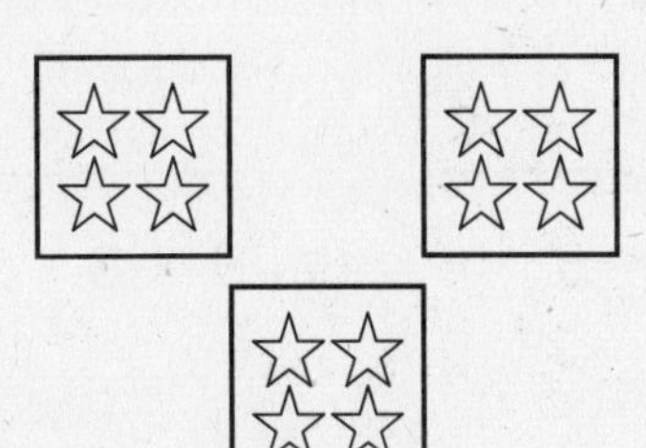

____

____

2.

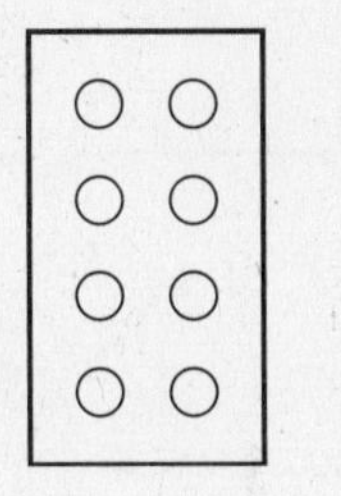

____

____

3.

____

____

Write the related addition or multiplication sentence.
Draw a picture that shows the sentence.

4. $3 + 3 + 3 + 3 = 12$

____

5. 2 groups of 5 equal 10.

____

6. $4 \times 2 = 8$

____

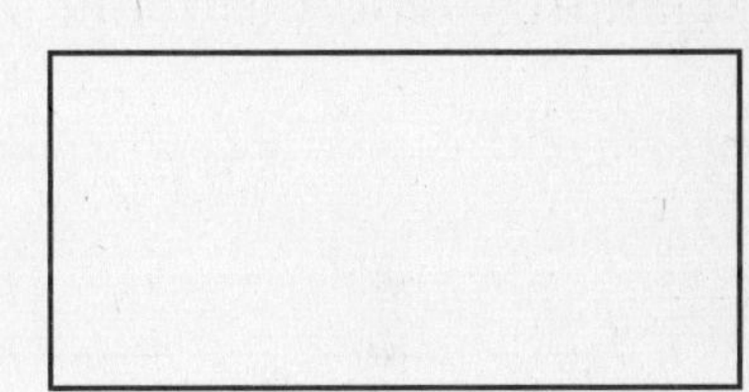

Problem Solving and TAKS Prep

7. Sue drives 36 miles to the fair. Every 4 miles, she sees a sign for the fair. How many signs will Sue see on her trip in all?

____

8. Mrs. Longo took 6 children to the fair. She bought each of them a stuffed animal that cost $5. How much did Mrs. Longo spend on stuffed animals?

____

9. An amusement park ride has 4 cars. Each car holds 4 people. How many people can go on the ride?

A 4

B 8

C 12

D 16

10. Each ticket at the fair costs $2. How much do 9 tickets cost?

F $2

G $9

H $11

J $18

Name____

Lesson 4.2

Multiply Facts Through 5

Find the product.

1. 4×8 ______
2. 3×7 ______
3. 2×9 ______
4. 5×7 ______
5. 0×2 ______
6. 3×9 ______
7. 4×1 ______
8. 5×8 ______
9. 5×5 ______
10. 3×8 ______
11. 3×3 ______
12. 2×8 ______
13. 4×7 ______
14. 5×3 ______
15. 2×2 ______
16. 5×0 ______
17. 2×1 ______
18. 5×4 ______
19. 4×4 ______
20. 3×6 ______

ALGEBRA Find the value of $3 \times n$ for each value of n.

21. $n = 3$ ______
22. $n = 2$ ______
23. $n = 0$ ______
24. $n = 5$ ______
25. $n = 4$ ______

Problem Solving and TAKS Prep

26. Adam buys 5 barbeque sandwiches at $5 each. How much money does Adam spend in all for the sandwiches?

____

27. The drink stand at the barbeque sells 4 drinks every minute. How many drinks will they sell in 9 minutes?

____

28. Jeff puts all of his toy cars in 3 rows. He puts 8 cars in each row. How many cars does Jeff have in all?

____

29. Brian makes 5 telephone calls every day for 9 days. How many telephone calls did Brian make in 9 days?

____

Name________________________________

Multiply Facts Through 10

Find the product. Show the strategy you used.

1. 8×8 ____________
2. 9×7 ____________
3. 6×9 ____________
4. 8×4 ____________

5. $\begin{array}{r} 8 \\ \times 3 \\ \hline \end{array}$
6. $\begin{array}{r} 7 \\ \times 7 \\ \hline \end{array}$
7. $\begin{array}{r} 10 \\ \times 7 \\ \hline \end{array}$
8. $\begin{array}{r} 8 \\ \times 6 \\ \hline \end{array}$

ALGEBRA Find the value of the coins.

9.

Nickels	1	3	5	7	9
Cents	5	■	■	■	■

____

10.

Dimes	1	2	6	9
Cents	10	■	■	■

____

Problem Solving and TAKS Prep

11. Six people can sit at a picnic table. How many people can sit at 8 picnic tables?

____

12. Jo has 9 dimes. She can buy 4 pencils with one dime. How many pencils can Jo buy?

____

13. Explain how to use a multiplication table to find 4×7.

____

____

____

____

____

14. Margo buys 6 apples each week. How many apples does Margo buy in 10 weeks?

A 6

B 10

C 16

D 60

Name ________________________________

Multiplication Table Through 12

Find the product. Show the strategy you used.

1. 11×2 ________

2. 8×11 ________

3. 7×12 ________

4. 9×12 ________

For 5–6, use the multiplication table.

5. What are the multiples of 6?

____

____

____

6. What are the multiples of 9?

____

____

____

×	0	1	2	3	4	5	6	7	8	9	10	11	12
0	0	0	0	0	0	0	0	0	0	0	0	0	0
1	0	1	2	3	4	5	6	7	8	9	10	11	12
2	0	2	4	6	8	10	12	14	16	18	20	22	24
3	0	3	6	9	12	15	18	21	24	27	30	33	36
4	0	4	8	12	16	20	24	28	32	36	40	44	48
5	0	5	10	15	20	25	30	35	40	45	50	55	60
6	0	6	12	18	24	30	36	42	48	54	60	66	72
7	0	7	14	21	28	35	42	49	56	63	70	77	84
8	0	8	16	24	32	40	48	56	64	72	80	88	96
9	0	9	18	27	36	45	54	63	72	81	90	99	108
10	0	10	20	30	40	50	60	70	80	90	100	110	120
11	0	11	22	33	44	55	66	77	88	99	110	121	132
12	0	12	24	36	48	60	72	84	96	108	120	132	144

ALGEBRA Use the rule to find the missing numbers.

7. Multiply the input by 9.

Input	Output
2	☐
4	☐
6	☐
8	☐

8. Multiply the input by 6.

Input	Output
7	☐
8	☐
☐	54
10	☐

9. Multiply the input by 10.

Input	Output
1	☐
☐	20
5	☐
7	☐

Name____ Lesson 4.5

Problem Solving Workshop Strategy: Guess and Check

Problem Solving Strategy Practice

Guess and check to solve.

1. Donna and Max painted a fence made of posts. Donna painted twice as many posts as Max did. Together, they painted 36 posts. How many posts did Max paint?

____

2. Erin buys her lunch every day. Together, her sandwich and her milk cost $6. Her sandwich costs twice as much as her milk. How much does each item cost?

____

3. A package contains one large sheet of stickers and one small sheet of stickers. The large sheet holds twice as many stickers as the small sheet. There are a total of 24 stickers in the package. How many stickers are on the small sheet?

____

Mixed Strategy Practice

4. **USE DATA** Ruby walks every day after school and records how far she walks. If the pattern continues, how many blocks will Ruby walk on Saturday? on Sunday?

Ruby's Walk Schedule

Day	Number of blocks
Monday	2
Tuesday	4
Wednesday	6
Thursday	8
Friday	10
Saturday	
Sunday	

____

5. Look at Exercise 2. Write a similar problem.

____

____

____

____

6. Mel is thinking of two odd numbers that add up to 14. What could Mel's numbers be?

____

Name____ Lesson 5.1

Algebra: Relate Multiplication and Division

Write the fact family for the set of numbers.

1. 4, 2, 8

2. 7, 2, 14

3. 8, 9, 72

4. 6, 1, 6

Find the value of the variable. Then write a related sentence.

5. $4 \times 7 = c$
$c =$ ____

6. $81 \div m = 9$
$m =$ ____

7. $16 \div j = 4$
$j =$ ____

8. $8 \times n = 16$
$n =$ ____

9. $64 \div 8 = r$
$r =$ ____

10. $7 \times 8 = w$
$w =$ ____

11. $9 \times 5 = p$
$p =$ ____

12. $10 \times 3 = a$
$a =$ ____

Problem Solving and TAKS Prep

13. Laura colors every picture in each of her 5 coloring books. There are 9 pictures in each book. How many pictures does Laura color in all?

14. Carlos has 63 crayons. He puts them into 7 equal groups for his classmates to use. How many crayons are in each group?

15. A crayon box holds 72 crayons. There are 9 equal rows of crayons in the box. How many crayons are in each row?

A 7
B 8
C 9
D 10

16. Mr. Lee draws a picture with 3 different crayons. A student uses 4 times as many crayons as Mr. Lee to draw another picture. How many crayons does the student use?

F 12
G 15
H 9
J 10

Name______________________________________

Divide Facts Through 5

Find the quotient.

1. $35 \div 5$ ______ **2.** $36 \div 4$ ______ **3.** $14 \div 2$ ______ **4.** $18 \div 3$ ______ **5.** $30 \div 3$ ______

6. $1 \div 1$ ______ **7.** $16 \div 4$ ______ **8.** $4 \div 4$ ______ **9.** $16 \div 2$ ______ **10.** $10 \div 2$ ______

11. $5\overline{)25}$ **12.** $3\overline{)6}$ **13.** $5\overline{)20}$ **14.** $5\overline{)45}$ **15.** $3\overline{)24}$

ALGEBRA Find the missing numbers.

16. $18 \div 2 = \square$ **17.** $15 \div 3 = \square$ **18.** $8 \div \square = 2$ **19.** $\square \div 2 = 5$

20. $32 \div 4 = \square$ **21.** $16 \div \square = 8$ **22.** $\square \div 1 = 3$ **23.** $12 \div 4 = \square$

Problem Solving and TAKS Prep

24. Ms. Jones will use vans to take her 45 students on a field trip. Each van will hold 9 students. How many vans will Ms. Jones need for the field trip?

____

25. A small airplane has a total of 27 seats. There are 3 seats in each row. How many rows are on the airplane?

____

26. Joanie divides a bag of 36 marbles evenly among 4 friends. How many marbles does each friend get?

A 36
B 9
C 8
D 4

27. It takes Annie 3 hours each time she mows the lawn. How many times can Annie mow the lawn in 12 hours?

F 12
G 6
H 4
J 3

Name______________________

Lesson 5.3

Divide Facts Through 10

Find the quotient. Show the strategy you used.

1. $48 \div 8$
2. $8\overline{)56}$
3. $10 \div 5$
4. $9\overline{)81}$

Write the related multiplication or division fact. Then find the quotient.

5. $64 \div 8$
6. $54 \div 9$
7. $72 \div 9$
8. $24 \div 6$

ALGEBRA Find the missing dividend or divisor.

9. $■ \div 6 = 3 + 4$

   $■ =$ ____

10. $45 \div ■ = 5 + 4$

    $■ =$ ____

11. $■ \div 6 = 5 + 1$

    $■ =$ ____

Problem Solving and TAKS Prep

12. Mike put 63 books away in the library. Only 7 books fit on each shelf. How many shelves did Mike use?

13. There are 9 oranges in a bowl. Sue, Jeff, and Tina take an equal number of oranges. How many did each person take?

14. Pam is 56 miles from Houston, Texas. If she sees a rest stop sign every 8 miles, how many signs will Pam see on the way to Houston?

    A 5
    C 7
    B 6
    D 8

15. Use a number line to find $60 \div 10$. Describe the pattern you used.

Name ____________________

Lesson 5.4

Divide Facts Through 12

Use the multiplication table to find the quotient. Write a related multiplication sentence.

1. $60 \div 12$

2. $90 \div 10$

3. $99 \div 11$

4. $96 \div 8$

5. $77 \div 7$

6. $121 \div 11$

7. $144 \div 12$

8. $90 \div 9$

×	0	1	2	3	4	5	6	7	8	9	10	11	12
0	0	0	0	0	0	0	0	0	0	0	0	0	0
1	0	1	2	3	4	5	6	7	8	9	10	11	12
2	0	2	4	6	8	10	12	14	16	18	20	22	24
3	0	3	6	9	12	15	18	21	24	27	30	33	36
4	0	4	8	12	16	20	24	28	32	36	40	44	48
5	0	5	10	15	20	25	30	35	40	45	50	55	60
6	0	6	12	18	24	30	36	42	48	54	60	66	72
7	0	7	14	21	28	35	42	49	56	63	70	77	84
8	0	8	16	24	32	40	48	56	64	72	80	88	96
9	0	9	18	27	36	45	54	63	72	81	90	99	108
10	0	10	20	30	40	50	60	70	80	90	100	110	120
11	0	11	22	33	44	55	66	77	88	99	110	121	132
12	0	12	24	36	48	60	72	84	96	108	120	132	144

ALGEBRA **Find the quotient or missing divisor.**

9. $84 \div 12 = ■$

■ = ____

10. $132 \div ■ = 12$

■ = ____

11. $72 \div ■ = 6$

■ = ____

12. $88 \div ■ = 11$

■ = ____

Problem Solving and TAKS Prep

13. Eli painted 11 stripes on each class flag he made. He painted 77 stripes in all. How many flags did Eli make?

14. There are 52 flags in the parade. The front row has 8 flags. Each of the other rows have 11 flags. How many rows have 11 flags?

15. Nick buys 12 boxes of crayons. There are 8 crayons in each box. How many crayons does Nick buy?

A 12

B 24

C 48

D 96

16. Which number is missing from the number sentence?

$66 \div ■ = 6$

F 7

G 9

H 11

J 13

Name ____________________

Problem Solving Workshop Skill: Choose the Operation

Problem Solving Skill Practice

Tell which operation you would use to solve the problem. Then solve the problem.

1. Sally takes 24 gallons of juice to the school picnic. The students at the picnic drink 2 gallons of juice every hour. How many hours will it take the students to drink all the juice?

____

2. Each student in Lori's class brings 12 cookies for the bake sale. There are 12 students in Lori's class. How many cookies does the class have for the bake sale?

____

Mixed Applications

3. Greg sells 108 mini muffins at the bake sale. He sold the mini muffins in bags of 12. How many bags of mini muffins does Greg sell? Write the fact family you used.

____

4. Julie wants to know how many workbooks she will use for the school year. The subjects she is studying are math, science, and reading. Each subject has 2 workbooks. Write a number sentence to show how many workbooks Julie will be using this year.

____

USE DATA Use the information in the table.

5. At the bake sale, 9 people buy slices of pie. Each person buys the same number of slices for $2 each. How many slices of pie does each person buy?

____

Bake Sale Final Sales	
cupcakes	147
cookies	211
slices of pie	54
slices of cake	39
brownies	97

6. How many cookies, brownies, and cupcakes were sold in all?

____

Name ______________________

Factors and Multiples

Use arrays to find all of the factors of each product. Write the factors.

1. 12 ______ ______

2. 18 ______ ______

3. 30 ______ ______

4. 21 ______ ______

List the first ten multiples of each number.

5. 11 ______ ______

6. 4 ______ ______

7. 9 ______ ______

8. 7 ______ ______

Is 8 a factor of each number? Write *yes* or *no*.

9. 16 ______

10. 35 ______

11. 56 ______

12. 96 ______

Is 32 a multiple of each number? Write *yes* or *no*.

13. 1 ______

14. 16 ______

15. 13 ______

16. 8 ______

Problem Solving and TAKS Prep

17. Tammy wants to make a pattern of multiples of 2, that are also factors of 16. What will be the numbers in Tammy's pattern?

____

18. Which multiples of 4 are also factors of 36?

____

19. Which multiple of 7 is a factor of 49?

A 1

B 4

C 7

D 9

20. Fred is placing 16 cups on a table, in equal rows. In what ways can he arrange these cups?

____

Name____ Lesson 6.1

Patterns on the Multiplication Table

Find the square number.

1. 9×9 ______

2. 5×5 ______

3. 10×10 ______

4. 4×4 ______

5. 2×2 ______

For 6–7, use the multiplication table.

6. What pattern do you see in the first 9 multiples of 11?

____

____

____

7. What pattern do you see in the first 9 multiples of 9?

____

____

____

×	0	1	2	3	4	5	6	7	8	9	10	11	12
0	0	0	0	0	0	0	0	0	0	0	0	0	0
1	0	1	2	3	4	5	6	7	8	9	10	11	12
2	0	2	4	6	8	10	12	14	16	18	20	22	24
3	0	3	6	9	12	15	18	21	24	27	30	33	36
4	0	4	8	12	16	20	24	28	32	36	40	44	48
5	0	5	10	15	20	25	30	35	40	45	50	55	60
6	0	6	12	18	24	30	36	42	48	54	60	66	72
7	0	7	14	21	28	35	42	49	56	63	70	77	84
8	0	8	16	24	32	40	48	56	64	72	80	88	96
9	0	9	18	27	36	45	54	63	72	81	90	99	108
10	0	10	20	30	40	50	60	70	80	90	100	110	120
11	0	11	22	33	44	55	66	77	88	99	110	121	132
12	0	12	24	36	48	60	72	84	96	108	120	132	144

Problem Solving and TAKS Prep

8. Niko has a square number that is less than 50. The digits add up to 9. What is Niko's number?

____

9. Use the rule *1 less than 3 times the number* to make a pattern. Start with 5. What is the 4th number in the pattern?

____

10. Which number has multiples with a repeating pattern of fives and zeros in the ones digit?

A 1

B 5

C 10

D 20

11. The multiples of which number are triple the multiples of 4?

F 8

G 12

H 40

J 84

Name________________________________

Lesson 6.2

Number Patterns

Find a rule. Then find the next two numbers in the pattern.

1. 108, 99, 90, 81, ■, ■

____

2. 2, 4, 6, 8, ■, ■

____

3. 2, 4, 8, 16, ■, ■

____

4. 85, 88, 82, 85, 79, 82, ■, ■

____

ALGEBRA Find a rule. Then find the missing numbers in the pattern.

5. 2, 6, 10, ■, 18, 22, 26, ■

____

6. 545, 540, 535, ■, 525, ■

____

7. 600, 590, 592, 582, 584, ■, ■

____

8. 400, 410, 409, ■, 418, ■

____

Use the rule to make a number pattern. Write the first four numbers in the pattern.

9. Rule: Add 7. Start with 14.

____

10. Rule: Subtract 6. Start with 72.

____

11. Rule: Add 2, subtract 5.
Start with 98.

____

12. Rule: Multiply by 2, subtract 1.
Start with 2.

____

Problem Solving and TAKS Prep

13. Look at the following number pattern. What is the next number, if the rule is multiply by 2?

3, 6, 12, □

14. Use the pattern 6, 9, 18, 21. What is a rule if the next number in this pattern is 42?

____

15. Which of the following describes a rule for this pattern?

3, 8, 5, 10, 7, 12

A Add 3, subtract 5

B Add 5, subtract 3

C Add 5, subtract 2

D Add 5, subtract 3

16. What might the next two numbers in this pattern be?

192, 96, 48, 24, □, □

Name____ Lesson 6.3

Problem Solving Workshop Strategy: Look for a Pattern

Problem Solving Strategy Practice

Find a pattern to solve.

1. A 3 by 3 array of blocks is painted so that every other row, starting with row 1, begins with a red block, and the alternate rows begin with a black block. Does the 12th row begin with red or black?

____

2. What will be the next three shapes in the pattern?

____

3. The first day on a March calendar is Saturday. March includes 31 days. On which day of the week will March end?

____

4. How many blocks are needed to build a stair step pattern that has a base of 10, a height of 10, and where each step is one block high and one block deep?

____

Mixed Strategy Practice

5. **USE DATA** If the pattern continues, how much would each 5-inch spike cost if you buy 10,000?

____

Ralph's Hardware Builder's Sale			
Spike	10	100	1,000
5-inch	10 cents ea.	8 cents ea.	6 cents ea.
10-inch	15 cents ea.	13 cents ea.	10 cents ea.
15-inch	20 cents ea.	16 cents ea.	12 cents ea.

6. Jules bought 5 pet turtles for $2 each. How much money did Jules spend on turtles in all?

____

7. Dorothy bought gloves with a $20 dollar bill. The gloves cost $6. How much change did Dorothy receive?

____

Name________________________________ Lesson 6.4

Multiplication Properties

Use the properties and mental math to find the value.

1. $3 \times 4 \times 2$ ______
2. $4 \times 5 \times 5$ ______
3. $7 \times 4 \times 0$ ______
4. $7 \times 12 \times 1$ ______

Find the missing number. Name the property you used.

5. $(5 \times 3) \times 4 = 5 \times (\blacksquare \times 4)$ ______
6. $3 \times 5 = 5 \times \blacksquare$ ______
7. $8 \times \blacksquare = (2 \times 10) + (6 \times 2)$ ______
8. $3 \times (7 - \blacksquare) = 3$ ______
9. $8 \times (5 - 3 - 2) = \blacksquare$ ______
10. $3 \times (2 \times 4) = \blacksquare \times (2 \times 3)$ ______

Make a model and use the Distributive Property to find the product.

11. 14×6 ______
12. 5×15 ______
13. 9×17 ______

Show two ways to group by using parentheses. Find the value.

14. $12 \times 5 \times 6$ ______ ______
15. $4 \times 3 \times 2$ ______ ______
16. $9 \times 3 \times 8$ ______ ______

Problem Solving and TAKS Prep

17. The pet store window has 5 kennels with 4 puppies in each and 6 kennels with 6 kittens in each. How many animals are in the window? ______

18. Jake takes his border collie on a walk for exercise. They walk four blocks that are 20 yards each. How many yards do Jake and his border collie walk? ______

19. Each packet of catnip toys has 7 toys. There are 20 packets in each box. How many toys are there in 5 boxes of catnip toys?

    A 500 C 700
    B 600 D 800

20. Is the number sentence true? $5 \times (4 - 3) = 5$ Explain. ______ ______

Name____ **Lesson 6.5**

Write and Evaluate Expressions

Write an expression that matches the words.

1. Stamps divided equally in 6 rows

____

2. Some peas in each of 10 pods

____

3. Some marbles on sale at 15¢ each

____

4. 1 pie cut into several equal slices

____

Find the value of the expression.

5. $y \times 5$ if $y = 6$ ________

6. $63 \div b$ if $b = 7$ ________

7. $9 \times a$ if $a = 2$ ________

8. $r \div 6$ if $r = 54$ ________

Match the expression with the words.

9. $(4 \times t) + 8$ ________

10. $(t \times 12) \div 4$ ________

11. $(t \div 2) - 8$ ________

a. a number, *t*, divided by 2 minus 8

b. 4 times a number, *t*, plus 8

c. a number, *t*, times 12 and separated into 4 groups

Problem Solving and TAKS Prep

12. Ella has some pages with 15 stickers on each page. Write an expression for the number of stickers Ella has.

____

13. Look at Exercise 12. Suppose Ella has 5 pages. How many stickers does she have in all?

____

14. Robert has 7 times as many soap box racers as Xavier. Let *r* represent the number of soap box racers Robert has. Which expression tells the number of racers Xavier has?

A $7 + r$

B $r - 7$

C $7 \times r$

D $r \div 7$

15. Fran needs to write 350 letters. She writes 35 letters each day. How may days does it take Fran to write all her letters? Explain.

____

____

Name ______________________

Lesson 6.6

Find Missing Factors

Find the missing factor.

1. $4 \times g = 20$

   $g =$ ____

2. $y \times 3 = 27$

   $y =$ ____

3. $8 \times w = 48$

   $w =$ ____

4. $7 \times a = 49$

   $a =$ ____

5. $\blacksquare \times 2 = 24$

   $\blacksquare =$ ____

6. $9 \times r = 81$

   $r =$ ____

7. $4 \times \blacksquare = 36$

   $\blacksquare =$ ____

8. $7 \times s = 77$

   $s =$ ____

9. $5 \times \blacksquare = 23 + 2$

   $\blacksquare =$ ____

10. $8 \times \blacksquare = 20 - 4$

    $\blacksquare =$ ____

11. $6 \times \blacksquare = 11 + 7$

    $\blacksquare =$ ____

12. $10 \times \blacksquare = 15 + 5$

    $\blacksquare =$ ____

13. $7 \times \blacksquare = 12 + 2$

    $\blacksquare =$ ____

14. $3 \times \blacksquare = 16 + 5$

    $\blacksquare =$ ____

15. $4 \times \blacksquare = 13 + 3$

    $\blacksquare =$ ____

Problem Solving and TAKS Prep

16. Each season, a total of 32 tickets are given away. Each chosen family is given 4 free tickets. Write a number sentence can be used to find the number of families that will receive tickets.

    ____

17. The manager of the Antelopes orders 4 uniforms for each new player. This year, the manager orders 16 uniforms. Write a number sentence that can be used to find the number of new players.

    ____

18. The Ants won 121 games this year. They win the same number of games for each of 11 months. Which number sentence can be used to find the number of games the Ants won each month.

    A $121 \times \blacksquare = 11$

    B $11 \times \blacksquare = 121$

    C $12 \times 12 = 144 + \blacksquare$

    D $10 \times \blacksquare = 100$

19. The community softball club has 120 members. They need to hire one coach for every 12 players. How many coaches will the community softball club need to hire?

    F 0

    G 11

    H 12

    J 10

Name________________________________

Multiplication and Division Equations

Write an equation for each. Choose the variable for the unknown. Tell what the variable represents.

1. Three students divide 27 bracelets equally among them.

____

2. Two pounds of beads put equally in bags makes a total of 50 pounds.

____

3. Maddie plants 3 seeds each in 15 pots.

____

4. Jesse divides 36 ornaments equally and puts them into 9 bags.

____

Solve the equation.

5. $a \times 6 = 48$

$a =$ _______

6. $d \div 4 = 7$

$d =$ _______

7. $3 \times w = 27$

$w =$ _______

8. $63 \div n = 9$

$n =$ _______

9. $b \div 5 = 5$

$b =$ _______

10. $22 \div t = 11$

$t =$ _______

11. $4 \times k \times 3 = 24$

$k =$ _______

12. $5 \times h \times 3 = 45$

$h =$ _______

Problem Solving and TAKS Prep

13. Phyllis is making rings. Each ring has 3 beads. If she can make 7 rings, how many beads does she have?

____

14. Ted divided 56 colored blocks into 8 bags. How many blocks were in each bag?

____

15. In which equation does $t = 3$?

A $t \div 12 = 4$

B $36 \div t = 12$

C $t \times 5 = 30$

D $15 \times t = 60$

16. Seven friends paid a total of $21 to enter a craft fair. Write an equation to show the price of one admission. Then solve the equation.

____

____

Name________________________________ Lesson 6.8

Problem Solving Workshop Strategy: Write an Equation

Problem Solving Strategy Practice

Write an equation to solve.

1. A scientist counted 9 sandhill crane sedges. Each had 7 sandhill crane nests with 2 eggs each. How many eggs did the scientist count in all?

____

2. When flamingos migrate, they do so at night. They can fly up to 600 km in one night. If their speed is about 60 kph, how many hours do they fly during the night?

____

3. Caribbean flamingos eat algae, insects, and small fish. They eat about 9 ounces of food a day. How many ounces of food do 5 flamingos eat in 14 days?

____

4. Flamingos pump water through their bills to filter out food. A Caribbean flamingo pumps water up to 300 times a minute. How many times does a flamingo pump water in a second?

____

Mixed Strategy Practice

5. One species of Sandhill cranes are 45 inches long. Another species, Lesser Sandhill cranes, are about 40 inches long. If 9 of each species stood end to end, how much shorter would the line of lesser sandhill cranes be?

____

6. A scientist was dyeing samples for testing. One by one, she dyed the samples in a sequence of green, yellow, red, blue, and pink. If the scientist continues with the pattern, what color will she dye sample 15?

____

7. **Pose A Problem** Use the information from exercise 5 to pose a new problem that can be solved by writing an equation.

____

____

____

____

8. **Write Math** In a particular type of bird the adult birds are either 7 or 8 inches long. A group of these birds stands end to end and the line is 64 inches long. What combinations of 7 and 8 inch birds could be in the line?

____

____

____

Name____

Ordered Pairs in a Table

Find a rule. Use your rule to find the missing numbers.

1.

Input, c	4	8	32	128	512
Output, d	1	2	8	☐	☐

____

2.

Input, r	4	5	6	7	8
Output, s	8	10	12	☐	☐

____

3.

Input, a	10	20	30	40	50
Output, b	1	2	3	☐	☐

____

4.

Input, m	85	80	75	70	65
Output, n	17	16	15	☐	☐

____

Use the rule to make an input/output table.

5. Multiply input by 7.

Input, a	1	2	3	4	5
Output, b	7	☐	☐	☐	☐

____

6. Divide input by 6.

Input, c	60	54	48	42	36
Output, d	10	☐	☐	☐	☐

____

Problem Solving and TAKS Prep

7. **USE DATA** Hal has 3 servings of milk each day. How many grams of protein will he get in 5, 6, and 7 days? Write an equation.

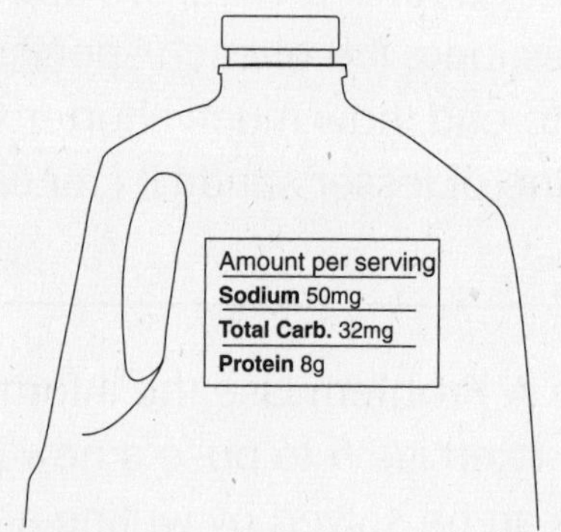

____

8. The table shows how many cups are in each pint. Which equation shows a rule for the table?

Pint, p	1	2	3	4	5
Cups, c	2	4	6	8	10

A $c \times 2 = p$

B $p \times 2 = c$

C $p \div 2 = c$

D $c + 2 = p$

9. The table shows how many glasses of water are in each pitcher. How many glasses of water are in 12 pitchers?

Pitchers, p	2	4	6	8	10
Glasses, g	6	12	18	24	30

____

Name ____________________

Lesson 7.1

Mental Math: Multiplication Patterns

Use mental math to complete the pattern.

1. $7 \times 6 = 42$

$7 \times 60 =$ ______

$7 \times 600 =$ ______

$7 \times 6{,}000 =$ ______

2. $3 \times 8 = 24$

$3 \times 80 =$ ______

$3 \times 800 =$ ______

$3 \times 8{,}000 =$ ______

3. $9 \times 7 = 63$

$9 \times 70 =$ ______

$9 \times 700 =$ ______

$9 \times 7{,}000 =$ ______

Use patterns and mental math to find the product.

4. 2×30 ______

5. 3×700 ______

6. $9 \times 4{,}000$ ______

7. 7×800 ______

ALGEBRA Find the value of *n*.

8. $2 \times n = 42{,}000$ $n =$ ______

9. $7 \times 400 = n$ $n =$ ______

10. $8 \times n = 16{,}000$ $n =$ ______

11. $n \times 500 = 4{,}500$ $n =$ ______

Problem Solving and TAKS Prep

12. Windsurfing costs $20 a day at New State Park. Jen windsurfed for 5 days. Paul windsurfed for 7 days. How much more did Paul pay than Jen?

______

13. Every carload of people entering the state park pays $7. In January, there were 200 cars that entered the park. In July, there were 2,000 cars that entered the park. How much more money did the park collect in July than in January?

______

14. Which number is missing from this equation?

$\blacksquare \times 7 = 3{,}500$

A 50
B 500
C 5,000
D 50,000

15. Which number is missing from this equation?

$8 \times \blacksquare = 32{,}000$

F 40
G 400
H 4,000
J 40,000

Name________________________________ Lesson 7.2

Mental Math: Estimate Products

Estimate the product. Write the method.

1. 2×49 ________ ________
2. 7×31 ________ ________
3. 5×58 ________ ________
4. 4×73 ________ ________
5. 3×27 ________ ________
6. 8×26 ________ ________
7. 4×25 ________ ________
8. 5×82 ________ ________
9. 6×53 ________ ________
10. 9×47 ________ ________
11. 6×71 ________ ________
12. 5×31 ________ ________
13. 88×2 ________
14. 29×8 ________
15. 65×4 ________
16. 39×7 ________

Problem Solving and TAKS Prep

USE DATA For 17–18, use the table.

17. About how many pencils will Haley use in 8 months?

____

18. How many more pencils will Haley use in ten months than Abby will use in ten months?

____

Pencils Used Each Month	
Name	**Number of Pencils**
Haley	18
Abby	12
Bridget	17
Kelsey	21

19. Which number sentence would give the best estimate of 6×17?

A 6×20
B 6×25
C 6×10
D 6×5

20. Which number sentence would give the best estimate of 6×51?

F 6×5
G 6×45
H 6×50
J 6×55

Name____ **Lesson 7.3**

Model 2-Digit by 1-Digit Multiplication

Find the product.

1.

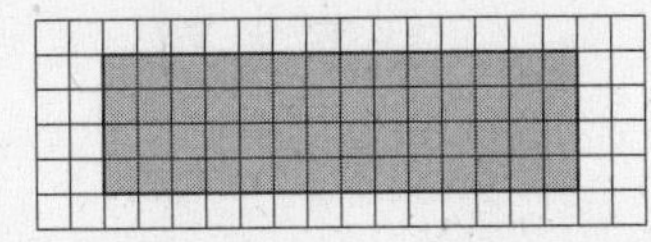

$4 \times 14 =$ ______

2.

$2 \times 13 =$ ______

3.

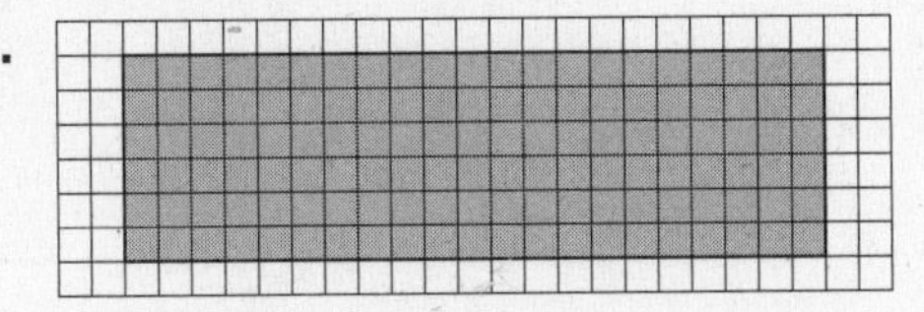

$6 \times 21 =$ ______

4.

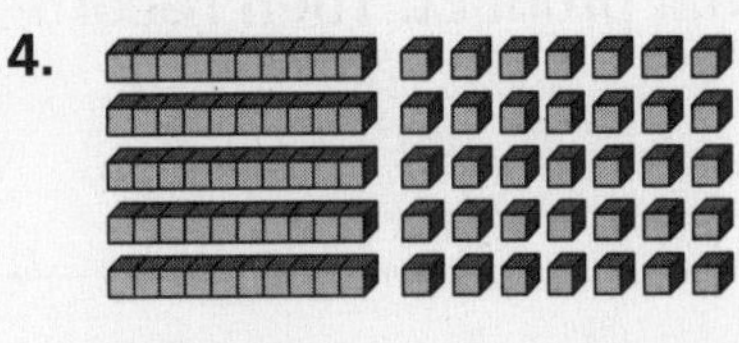

$5 \times 17 =$ ______

5.

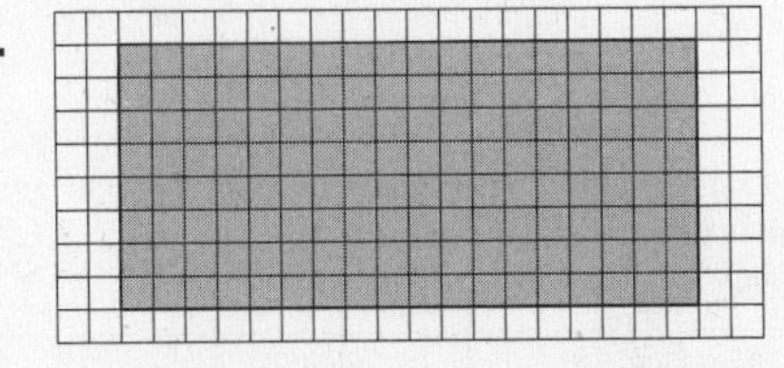

$8 \times 18 =$ ______

6.

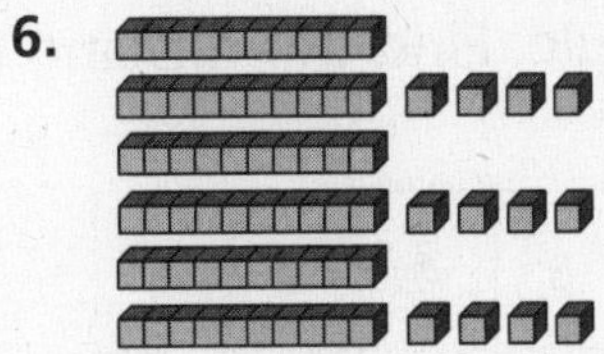

$3 \times 24 =$ ______

7.

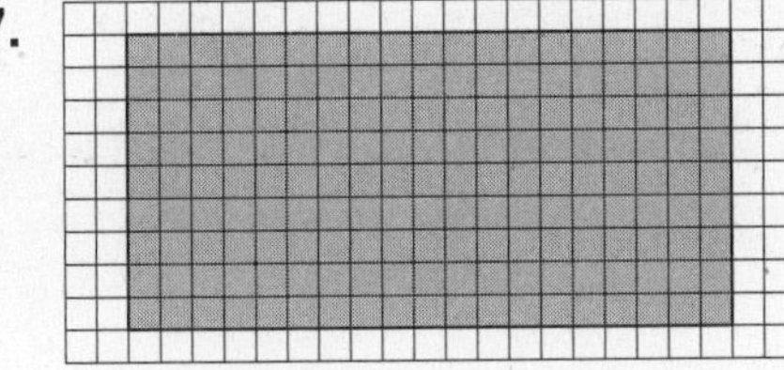

$9 \times 19 =$ ______

8.

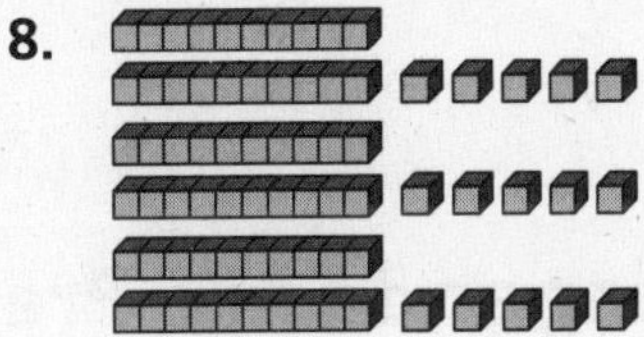

$3 \times 25 =$ ______

9.

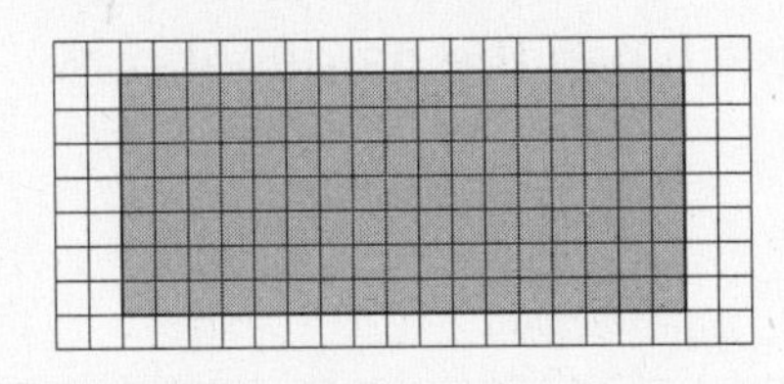

$7 \times 17 =$ ______

Use grid paper or base-ten blocks to model the product. Record your answer.

10. 2×18 ______

11. 5×16 ______

12. 4×17 ______

13. 3×31 ______

14. 6×17 ______

15. 8×18 ______

16. 7×31 ______

17. 9×33 ______

Name________________________________

Lesson 7.4

Record 2-Digit by 1-Digit Multiplication

Estimate. Then record the product.

1. $\begin{array}{r} 53 \\ \times 5 \\ \hline \end{array}$

2. $\begin{array}{r} 34 \\ \times 3 \\ \hline \end{array}$

3. $2 \times \$49$

______

4. 6×71

______

Write each partial product. Then record the product.

5. 9×62

______

______

6. 3×41

______

______

7. 5×38

______

______

8. 2×49

______

______

ALGEBRA Write a rule. Find the missing numbers.

9.

Number of quarts, (q)	2	3	4	5	6
Number of cups, (c)	8	12	■	20	■

Rule: ______

10.

Number of yards, (y)	1	2	3	4	5
Number of feet, (f)	3	■	9	12	■

Rule: ______

Problem Solving and TAKS Prep

11. Mr. Lewis gets in his car at 11:15 A.M. He drives for 2 hours and 45 minutes. What time is it when Mr. Lewis stops driving?

______

12. A gardener had 21 tulip bulbs. She bought 7 more. How many rows of 4 tulips each can the gardener now plant?

______

13. Jan and Beth count their savings. Jan has 7 one-dollar bills and one five-dollar bill. Beth has 3 ten-dollar bills. How much more do they need to save to have $50 total?

   A $11

   B $42

   C $8

   D $39

14. If the pattern below continues, could 180 be one of the products in this pattern? Explain.

   $3 \times 3 = 9$

   $3 \times 3 \times 3 = 21$

   $3 \times 3 \times 3 \times 3 = 81$

______

Name____ Lesson 7.5

Problem Solving Workshop Strategy: Draw a Diagram

Problem Solving Strategy Practice

Draw a diagram to solve.

1. Jan walks 5 blocks north, 1 block east, and 3 more blocks north. Then she walks 1 block west and 1 block south. How far is Jan from where she started?

____

2. Nick's toy boat is 24 inches long. Ben has 10 toy boats, but they are each only 6 inches long. How many of Ben's boats, laid end to end, would it take to match the length of Nick's boat?

____

Mixed Strategy Practice

USE DATA For 3–6, use the information in the table.

3. How many times greater is the maximum lifespan of 6 Bowhead Whales than that of 1 Fin Whale?

____

Whales' Maximum Life Span

Whale Type	Years
Pilot	60
Orca	90
Fin	60
Blue	80
Bowhead	130

4. List the types of whales shown in order from shortest lifespan to longest lifespan.

____

____

5. Look at Exercise 3. Write a similar problem using two different types of whales.

____

____

____

6. Write three different expressions that equal the life span of the Bowhead whale, using one or more operations.

____

____

____

Name ______________________________ Lesson 8.1

Multiply 2-Digit Numbers

Estimate. Then find the product.

1. $\begin{array}{r} 36 \\ \times 2 \\ \hline \end{array}$ ________

2. $\begin{array}{r} 99 \\ \times 3 \\ \hline \end{array}$ ________

3. $\begin{array}{r} 48 \\ \times 7 \\ \hline \end{array}$ ________

4. $\begin{array}{r} 19 \\ \times 9 \\ \hline \end{array}$ ________

5. 28×6 ________ ________

6. 52×4 ________ ________

7. 63×5 ________ ________

8. 72×8 ________ ________

ALGEBRA Find the missing digit.

9. $4 \times 27 = \blacksquare 08$

10. $\blacksquare \times 19 = 95$

11. $7 \times 2\blacksquare = 154$

12. $8 \times 7\blacksquare = 568$

13. $6 \times 47 = 2\blacksquare 2$

14. $2 \times \blacksquare 8 = 176$

15. $4 \times 2\blacksquare = 112$

16. $\blacksquare \times 63 = 189$

17. $9 \times 97 = 8\blacksquare 3$

Problem Solving and TAKS Prep

18. There are 7 baseball teams that played 18 games each. How many games did the teams play in all?

____

19. Joe's team scores 6 goals each game. How many did Joe's team score in 18 games?

____

20. Bea bought 2 packages of smiley-face stickers. She bought 49 stickers in all. One package had 3 more stickers than the other. How many stickers were in each package?

A 12, 9
B 49, 49
C 23, 26
D 14, 32

21. Gene baked 9 trays of chicken for a cookout. Each tray had 16 pieces of chicken. How many pieces of chicken did Gene bake in all?

F 90
G 94
H 104
J 144

Name____________________

Lesson 8.2

Multiply 3-Digit Numbers

Estimate. Then find the product.

1. $\begin{array}{r} 271 \\ \times\ 4 \\ \hline \end{array}$ ______

2. $\begin{array}{r} 435 \\ \times\ 6 \\ \hline \end{array}$ ______

3. $\begin{array}{r} 825 \\ \times\ 5 \\ \hline \end{array}$ ______

4. $\begin{array}{r} 681 \\ \times\ 8 \\ \hline \end{array}$ ______

5. 547×3 ______

6. 223×7 ______

7. 424×9 ______

8. 999×2 ______

ALGEBRA **Find the missing numbers.**

9. $\blacksquare \times 131 = 524$

10. $7 \times \blacksquare 52 = 3,164$

11. $8 \times 65\blacksquare = 5,224$

12. $\blacksquare \times 7\blacksquare 9 = 5,992$

13. $3 \times \blacksquare 24 = 1,872$

14. $5 \times \blacksquare 5 \blacksquare = 3,765$

Problem Solving and TAKS Prep

15. Horace's recipe makes 48 biscuits in each batch. He makes 7 batches for a community bazaar. How many biscuits does Horace bake in all?

______

16. Janice bakes 35 casseroles a day at her restaurant. How many casseroles does Janice bake in 8 days?

______

17. Elle's box holds 175 paper clips. How many paper clips are in 8 boxes?

A 800
B 900
C 1,360
D 1,400

18. Tim's drawer holds 125 trading cards. He has 6 drawers full of trading cards. How many cards does Tim have in all?

F 250
G 600
H 1,250
J 750

Name________________________________

Multiply with Zeros

Estimate. Then find the product.

1. 3×304 **2.** 5×470 **3.** 6×705 **4.** $4 \times \$430$

______ ______ ______ ______

5. 2×807 **6.** 7×130 **7.** $6 \times \$304$ **8.** 8×510

______ ______ ______ ______

ALGEBRA **Find the value of the expression** $n \times 106$ **for each value of** n**.**

9. $\$106 \times 3$ **10.** $\$106 \times 5$ **11.** $\$106 \times 2$ **12.** $\$106 \times 9$

Problem Solving and TAKS Prep

13. It is 18 miles from Saya's house to the place where she takes her computer lessons. Saya drives there and back each day for 8 days. How far does she drive in all?

______

14. Loren built a model tree house that is 105 inches tall. The real tree house will be 3 times as tall. How tall will the real tree house be?

______

15. Mr. Bench bought 4 pairs of pajamas for his children for $20 each. How much did Mr. Bench spend in all?

A $75
B $78
C $80
D $85

16. Arthur buys 6 new shirts for $10 each. How much does Arthur spend in all?

F $55
G $66
H $60
J $75

Name ____________________

Choose a Method

Find the product. Write the method you used.

1. $1{,}000 \times 8$ = ________

2. 322×5 = ________

3. $2{,}168 \times 4$ = ________

4. $4{,}422 \times 2$ = ________

5. $2{,}121 \times 6$ = ________

6. 500×7 = ________

7. $6{,}797 \times 3$ = ________

8. $9{,}009 \times 9$ = ________

9. 604×8 = ________

10. 550×5 = ________

11. 667×6 = ________

12. $3{,}923 \times 1$ = ________

ALGEBRA Find the missing number.

13. $\blacksquare \times 749 = 2{,}247$

14. $5 \times 612 = 3{,}\blacksquare 60$

15. $8 \times 3\blacksquare 2 = 2{,}816$

16. $6 \times 434 = 2{,}\blacksquare 04$

17. $4 \times 35\blacksquare = 1{,}432$

18. $7 \times 635 = \blacksquare{,}445$

Problem Solving and TAKS Prep

19. There are 8 campers getting ready to camp out for one week. Each camper packs 21 meals for the campout. How many meals do the campers pack in all?

____

20. Some campers plan to bring 3 water bottles per person per day on their camping trip. How many water bottles will 8 campers need for 4 days of camping?

____

21. What is the product of 4×100? Explain how you solved the problem.

____

____

____

____

22. What is the product of 6×205? Explain how you solved the problem.

____

____

____

____

Name____

Problem Solving Workshop Skill: Evaluate Reasonableness

Problem Solving Skill Practice

Solve the problem. Then evaluate the reasonableness of your answer.

1. Mr. Kohfeld buys a $1.37 carton of eggs each week. How much does Mr. Kohfeld spend on eggs in 4 weeks?

____

2. Vivian spends $6.49 on lunch every day. How much does Vivian spend on lunch in 7 days?

____

3. Yoshi is an athlete who eats a breakfast of 1,049 calories each morning. How many calories does Yoshi consume at breakfast in 7 days?

____

4. Together Elise and Chris spelled 27 words correctly. Chris spelled 5 more words correctly than Elise. How many words did each student spell correctly?

____

Mixed Applications

Solve.

5. The Miller family eats 9 bowls of cereal each day. How many bowls of cereal does the Miller family eat in a year (365 days)? How do you know your answer is reasonable?

____

6. Joe spent $25.87 for groceries. He bought cereal for $6.25, eggs for $5.37, pancake mix for $3.67, bacon for $7.25, and juice. How much did Joe spend on juice?

____

USE DATA For 7–8, use the information in the stone wall.

7. Tanya is building this wall from stone. If the pattern continues, how thick will the next stone be?

____

8. If the finished wall is 6 stones high overall what is the finished height of the wall?

____

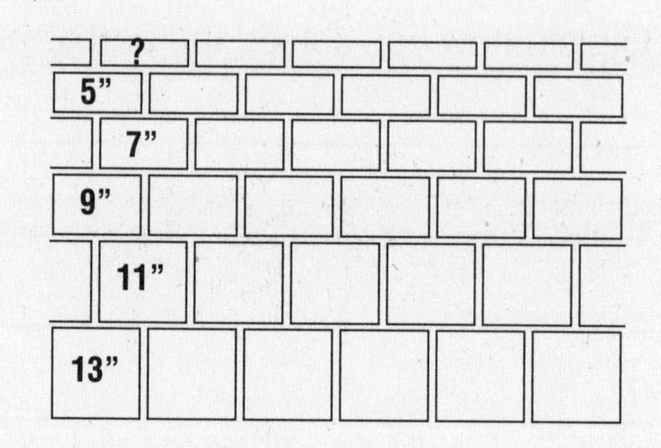

Name________________________________ Lesson 9.1

Mental Math: Multiplication Patterns

Use mental math and patterns to find the product.

1. 50 × 3,000 ____________

2. 7 × 40 ____________

3. 8 × 1,000 ____________

4. 50 × 700 ____________

5. 12 × 2,000 ____________

6. 70 × 200 ____________

7. 11 × 120 ____________

8. 90 × 80 ____________

ALGEBRA **Copy and complete the tables using mental math.**

9. 1 roll = 20 nickels

Number of rolls	20	30	40	50	600
Number of Nickles	400	■	■	■	■

10. 1 roll = 60 dimes

Number of rolls	20	30	40	50	600
Number of Dimes	1,200	■	■	■	■

	x	7	60	700	8,000
11.	40	2,80	■	■	■
12.	60	■	■	■	480,000

	x	8	40	500	9,000
13.	50	400	■	■	■
14.	90	■	■	■	810,000

Problem Solving and TAKS Prep

USE SCIENCE DATA **For 15–16, use the table.**

15. How long would a drywood termite magnified by 6,000 appear to be?

____

16. Which would appear longer, a drywood termite magnified 1,200 times or a wasp magnified 900 times?

____

Insect Lengths	
Insect	**Length (in mm)**
Carpenter Bee	19
Drywood Termite	12
Fire Ant	4
Termite	12
Wasp	15

17. How many zeros are in the product of 400 × 500?

A 4

B 5

C 6

D 7

18. How many zeros must be in the product of 1,000 and any factor?

____

Name________________________________

Multiply by Tens

Choose a method. Then find the product.

1. 20×17 ____
2. 15×60 ____
3. 66×50 ____
4. 78×30 ____
5. 96×40 ____
6. 90×46 ____
7. 52×80 ____
8. 70×29 ____

ALGEBRA Find the missing digit.

9. $22 \times 3\blacksquare = 660$ ____
10. $60 \times 37 = 2{,}\blacksquare20$ ____
11. $5\blacksquare \times 80 = 4{,}480$ ____
12. $\blacksquare0 \times 77 = 3{,}080$ ____
13. $40 \times 44 = \blacksquare{,}760$ ____
14. $90 \times 83 = 7{,}4\blacksquare0$ ____

Problem Solving and TAKS Prep

USE DATA For 15–17, use the table.

Animated Productions		
Title	Date Released	Frames Per Second
The Enchanted Drawing ©	1900	20
Little Nemo ©	1911	16
Snow White and the Seven Dwarfs ©	1937	24
Pinocchio ©	1940	19
The Flintstones TM	1960-1966	24

15. How many frames does it take to produce 60 seconds of Snow White?

____

16. Are there more frames in 30 seconds of Pinocchio or 45 seconds of The Enchanted Drawing?

____

17. Sadie runs 26 miles each week. How many miles will Sadie run in 30 weeks?

 A 780
 B 720
 C 690
 D 700

18. If gourmet cookies cost $12 a pound, how much does it cost to purchase 30 pounds of cookies?

 F $360
 G $3,600
 H $36
 J $36,000

Name____________________

Mental Math: Estimate Products

Choose the method. Estimate the product.

1. 34×34
2. 27×42
3. 41×55
4. 17×39

____ ____ ____ ____

5. 72×21
6. 54×67
7. 58×49
8. 64×122

____ ____ ____ ____

9. 93×93
10. 19×938
11. 42×666
12. 71×488

____ ____ ____ ____

Problem Solving and TAKS Prep

13. **Fast Fact** A serving of watermelon has 27 grams of carbohydrate. About how many grams of carbohydrate do 33 servings contain?

____

14. There are 52 homes in Ku's neighborhood. If the door on each refrigerator in each home is opened 266 times a week, and each home has one refrigerator, about how many times are the doors opened in all?

____

15. Choose the best estimate for the product of 48×637.

A 20,000
B 24,000
C 30,000
D 34,000

16. An assembly line produces enough cotton for 1,500 T-shirts a day. How could you estimate the number of T-shirts 45 assembly lines produce?

F $1{,}500 \times 50$
G $30 \times 1{,}200$
H $2{,}000 \times 100$
J $150 \times 4{,}500$

Name____________________

Problem Solving Workshop Strategy: Solve a Simpler Problem

Problem Solving Strategy Practice

Solve a simpler problem.

1. For a year, Greta counted the wrens in her backyard. She counted an average of 20 each day. About how many wrens did Greta count in all over the year?

____

2. Participants in the Backyard Birdcount reported seeing 843,635 Canadian geese and 710,337 snow geese. How many more Canadian geese were counted?

____

3. Participants reported seeing 486,577 European starlings and 254,731 American robins. How many starlings and robins were reported in all?

____

4. The 93 counties in Nebraska reported seeing an average of 5,245 birds. About how many birds were reported in Nebraska?

____

Mixed Strategy Practice

USE DATA For 5–6, use the table.

Type of Bird	Maximum Speed (in miles per hour)	Maximum Length (in inches)
Carrion Crow	31	20
House Sparrow	31	6
Mallard	41	26
Wandering Albatross	34	48

5. Participants in Lincoln, Nebraska, reported seeing 311,214 birds while those in Hutchinson, Kansas, reported seeing 133,288. How many birds did the locations report in all?

____

6. Jim saw a hummingbird that was only 4 inches long. Which bird has a length that is 5 times as long?

____

7. Pose a Problem Look back at Exercise 4. Write a similar problem by changing the number of counties and the average number of birds seen.

____

____

____

Model 2-Digit by 2-Digit Multiplication

Use the model and partial products to solve.

1. 15×29

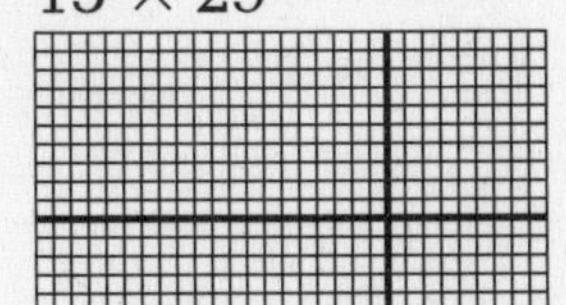

______

2. 17×32

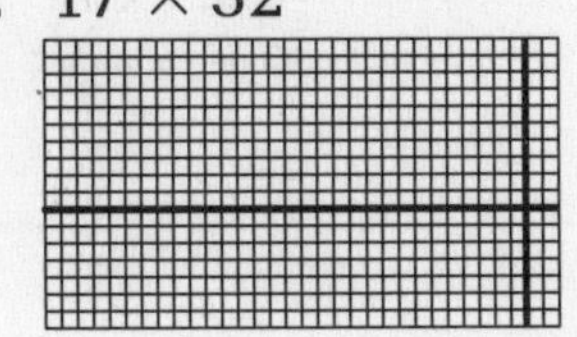

______

3. 19×25

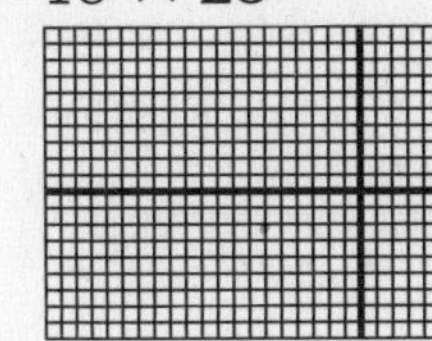

______

4. 14×27

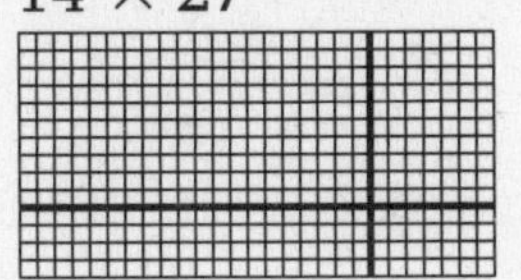

______

5. 16×28

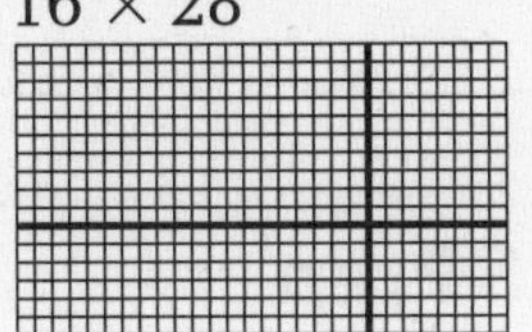

______

6. 19×24

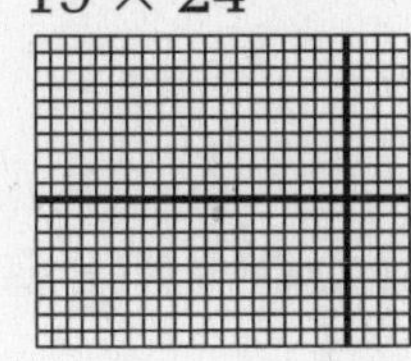

______

7. 17×26

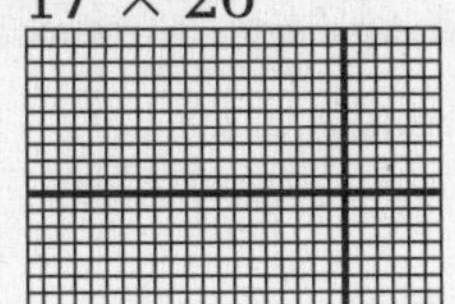

______

8. 18×21

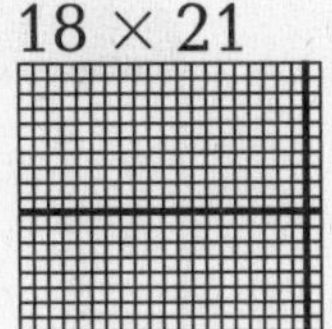

______

9. 26×36

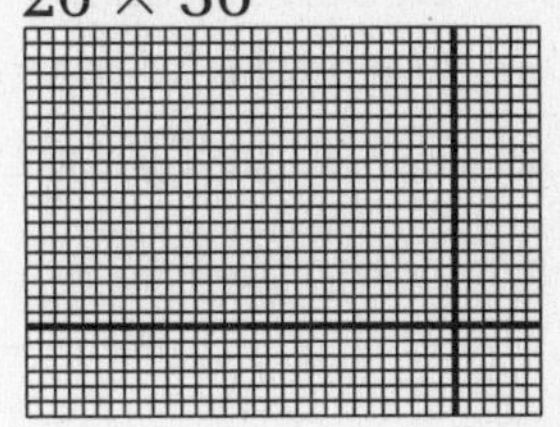

______

Problem Solving and TAKS Prep

10. The apples from an average tree will fill 20 bushel-sized baskets. If an orchard has 17 average trees, how many baskets of apples can it produce?

______

11. If each student eats about 65 apples a year, how many apples will the 27 students in Mrs. Jacob's class eat in all?

______

12. Draw a model in the space below that could represent the product 64.

______

13. What product is shown by the model?

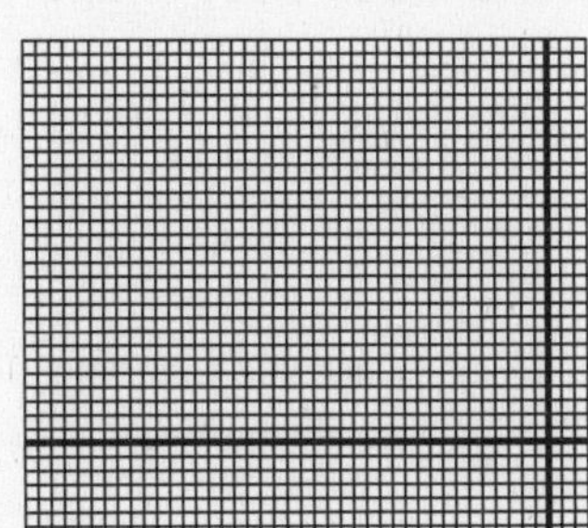

______

Name_______________________________________

Record 2-Digit by 2-Digit Multiplication

Estimate. Then choose either method to find the product.

1. 28×19 ________

2. 36×53 ________

3. $\$76 \times 25$ ________

4. 64×31 ________

5. 76×83 ________

6. 41×69 ________

7. 57×65 ________

8. $82 \times \$48$ ________

Problem Solving and TAKS Prep

USE DATA For 9–10, use the bar graph.

9. Sun Beach Parasail had 19 riders each windy day. How many riders in all parasailed last year on windy days?

____

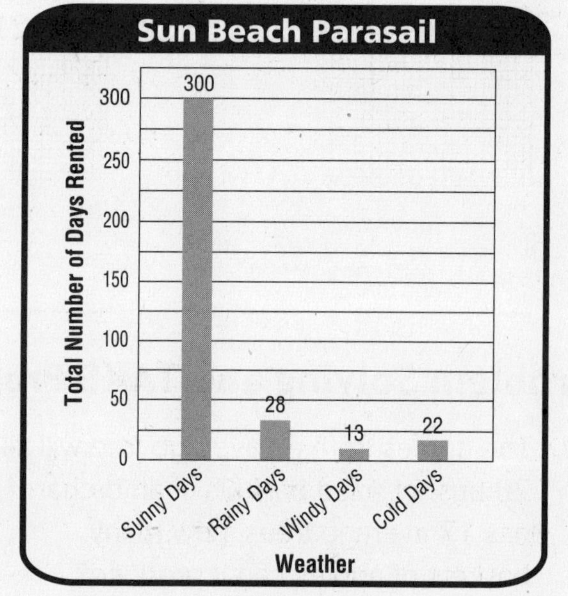

10. On each of 75 sunny days, Sun Beach Parasail had 62 riders. How many riders in all parasailed on those 75 days?

____

11. Willa bought 16 arborvitae trees for her backyard. Each tree cost \$33. How much did the trees cost in all?

A \$300
B \$480
C \$528
D \$600

12. There are 47 members in the Fun in the Sun Parasail Club. Each member spent 88 hours last year parasailing. How many hours did the club members spend parasailing last year in all?

F 6,413
G 4,136
H 4,230
J 7,236

Name________________________________

Practice Multiplication

Estimate. Find the product.

1. 58×39 ______
2. 48×45 ______
3. 62×76 ______
4. 19×37 ______
5. 97×36 ______
6. 54×47 ______
7. 37×68 ______
8. 77×23 ______
9. 24×42 ______
10. 37×19 ______
11. 88×63 ______
12. 13×57 ______

Problem Solving and TAKS Prep

13. For a field trip, 11 school buses take students from school to the car museum. There are 42 students on each bus, how many students go to the car museum in all?

______

14. At the car museum 12 groups of 35 students go into the engine history room. How many students go into the engine history room in all?

______

15. Kip likes multigrain bread that costs $15 for 3 loaves. If his family consumes 3 loaves every week, how much will they spend on this bread in one year?

   A $780
   B $790
   C $795
   D $800

16. A tee shirt company ordered 25 boxes of plain tee shirts. Each box holds 110 tee shirts. How many tee shirts did the company order?

   F 1,150
   G 2,110
   H 2,500
   J 2,750

Name______________________________

Choose a Method

Estimate. Find the product. Write the method you used.

1. $\begin{array}{r} 22 \\ \times 30 \\ \hline \end{array}$ ____________

2. $\begin{array}{r} 653 \\ \times 31 \\ \hline \end{array}$ ____________

3. $\begin{array}{r} 500 \\ \times 70 \\ \hline \end{array}$ ____________

4. $\begin{array}{r} 322 \\ \times 23 \\ \hline \end{array}$ ____________

5. $\begin{array}{r} 312 \\ \times 20 \\ \hline \end{array}$ ____________

6. $\begin{array}{r} 666 \\ \times 11 \\ \hline \end{array}$ ____________

7. $\begin{array}{r} 87 \\ \times 59 \\ \hline \end{array}$ ____________

8. $\begin{array}{r} 900 \\ \times 80 \\ \hline \end{array}$ ____________

9. 343×22 ____________

10. 505×90 ____________

11. 62×27 ____________

12. 52×75 ____________

ALGEBRA Use a calculator to find the missing digit.

13. $67 \times 457 = 30,\blacksquare 19$ ____________

14. $\blacksquare 4 \times 367 = 30,828$ ____________

15. $38 \times 2\blacksquare 9 = 9,082$ ____________

Problem Solving and TAKS Prep

16. June had a party at home. The special plates cost $13 each. If there were a total of 23 people at the party, how much did June spend on plates?

____

17. A local store sells silver balloons at $29 a case. Frank bought 48 cases. How much did the balloons cost?

____

18. What is the best method to multiply 40×800?

    A mental math

    B calculator

    C paper and pencil

    D none of the above

19. What is the best method to multiply $\$27 \times 54$?

    F mental math

    G calculator

    H paper and pencil

    J none of the above

Name____ Lesson 9.9

Problem Solving Workshop Skill: Multistep Problems

Problem Solving Skill Practice

1. The Pacific Wheel is a ferris wheel that can carry 6 passengers in each of 20 cars in one ride. How many passengers can it carry on a total of 45 rides?

____

2. Bus A travels 532 miles one way. Bus B travels 1,268 miles round trip. Which bus travels the most round-trip miles if Bus A makes 6 trips and Bus B makes 5 trips?

____

3. There are 62 students in all. Twenty-five take only band class. Thirty-four take only art class. The rest take both band and art class. How many students take both band and art?

____

4. Trin bought 6 tee-shirts for $17 each. Ron bought 7 shirts at the same price. How much Trin and Ron spend altogether?

____

Mixed Applications

USE DATA For 5–6, use the table.

5. **USE DATA** The Smiths are a family of 7. How much will they spend for admission to the carnival if they go on Saturday night?

____

6. **USE DATA** How much will the Smiths save if they go on Monday instead of Saturday?

____

Carnival One Night Admission Tickets

Night	Cost
Monday through Wednesday	$12
Thursday through Friday	$15
Saturday	$20

7. A local carnival has a Ferris wheel with 20 cars that seat 4 people each. Each ride is 10 minutes with 5 minutes to unload and reload. How many people can the Ferris wheel carry in 3 hours?

____

8. Rosa rode the Ferris wheel, the go-carts for 10 minutes, the merry-go-round for 25 minutes, and the roller coaster for 35 minutes. She was on rides for 1 hour and 30 minutes. How long did she ride the Ferris wheel?

____

Name ____________________

Lesson 10.1

Points, Lines, and Rays

Name the geometric term that best represents the object.

1. top of a desk ____________

2. chalk tray ____________

3. a point from Earth into space ____________

4. NNE on a compass ____________

Name an everyday object that represents the term.

5. point ____________

6. ray ____________

7. line segment ____________

8. plane ____________

Draw and label an example of each on the dot paper.

9. plane *ABC*

10. line segment *DE*

11. ray *FG*

12. point *H*

Problem Solving and TAKS Prep

USE DATA For 13–16, use the picture of a hallway.

13. What geometric term describes where the ceiling meets a wall?

____

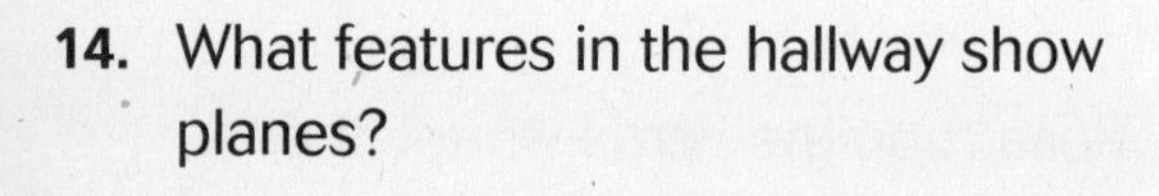

14. What features in the hallway show planes?

____

15. What geometric term best describes the arrow?

A line
B line segment
C point
D ray

16. Which geometric term best describes the black dot on the window?

F line
G line segment
H point
J ray

Name________________________________ **Lesson 10.2**

Classify Angles

Classify each angle as *acute, right,* or *obtuse.*

1.

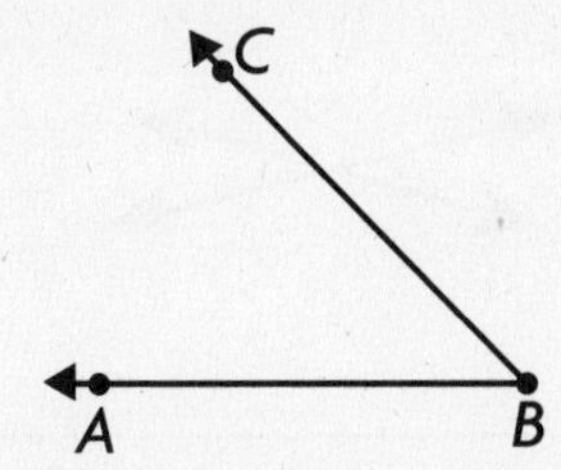

____

2.

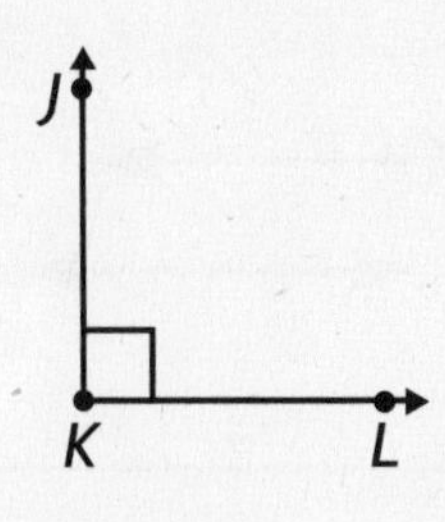

____

3.

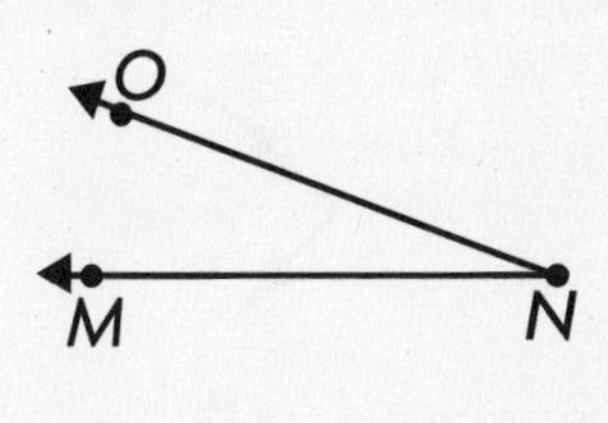

____

Draw and label an example of each.

4. acute angle *PQR*

5. obtuse angle *STU*

6. right angle *DEF*

7. acute angle *XYZ*

8. obtuse angle *JKL*

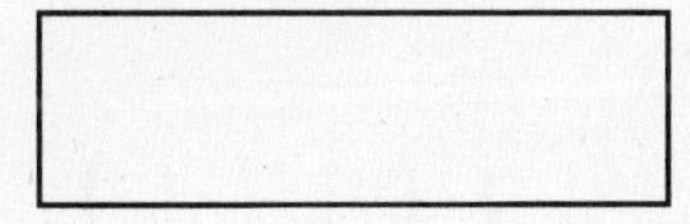

9. right angle *GHI*

Problem Solving and TAKS Prep

USE DATA For 10–11, use the angles shown at the right.

10. Which angles appear to be acute?

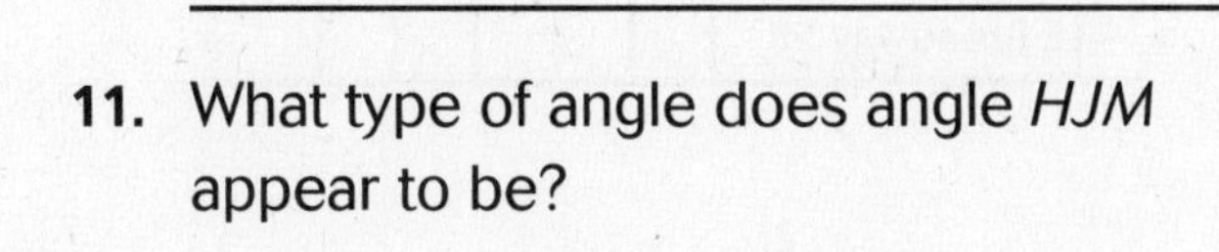

____

11. What type of angle does angle *HJM* appear to be?

____

G K H J L M

12. At what time do the hands on a clock represent a right angle?

A 9:15 **C** 9:00

B 11:30 **D** 6:00

13. Which is the measure of a right angle?

F 45° **H** 110°

G 90° **J** 180°

Name______________________________ Lesson 10.3

Line Relationships

Name any line relationships you see in each figure. Write *intersecting*, *parallel*, or *perpendicular*.

1. 

____

2. 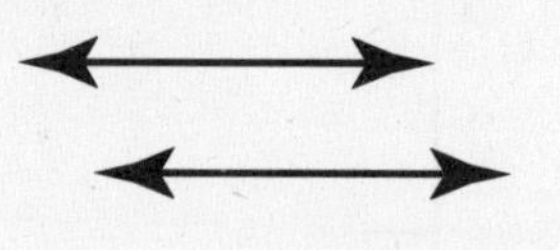

____

3. 

____

4. 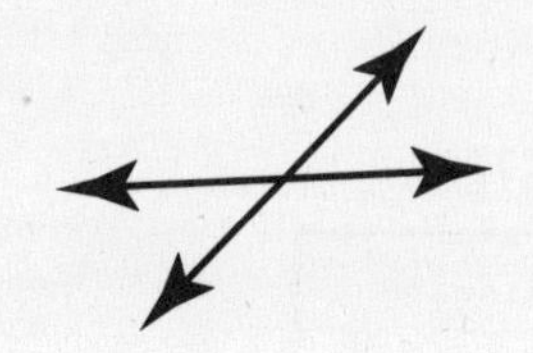

____

____

5. 

____

____

6. 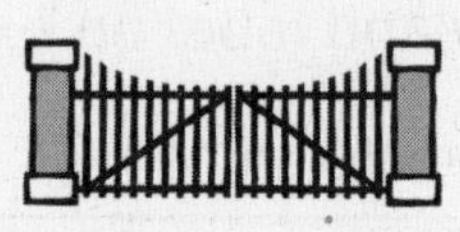

____

____

Problem Solving and TAKS Prep

USE DATA For 7–8, use the map.

7. Name a street that appears to be parallel to E Broadway St.

____

8. Name a street that intersects NE Madison St and appears to be parallel to 15th Ave NE.

____

9. Which best describes intersecting lines?

A They never meet.

B They form four angles.

C They form only obtuse angles.

D They form only acute angles.

10. Which best describes parallel lines?

F They never meet.

G They form four angles.

H They form only obtuse angles.

J They form only acute angles.

Name______________________________________

Polygons

Name the polygon. Tell if it appears to be *regular* or *not regular*.

1. 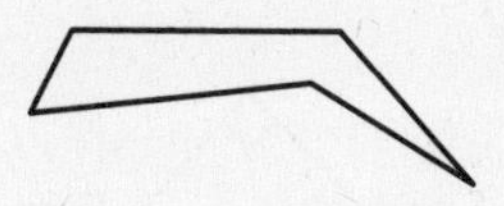

____

2. 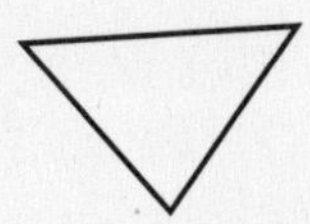

____

3. 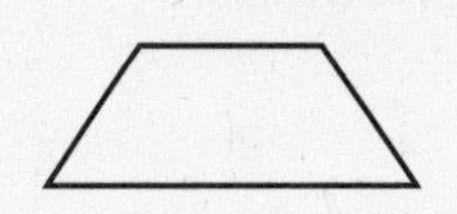

____

4. 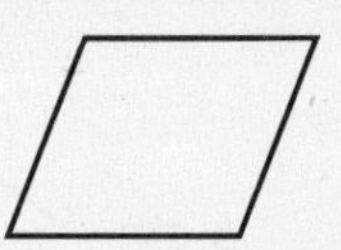

____

Tell if each figure is a polygon. Write *yes* or *no*. Explain.

5. 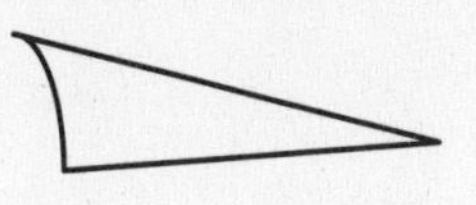

____

6. 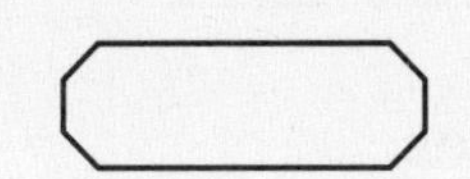

____

7. 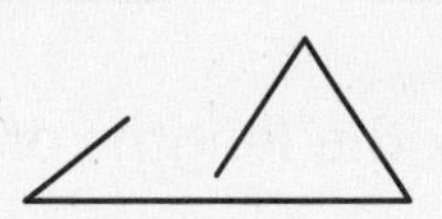

____

8. 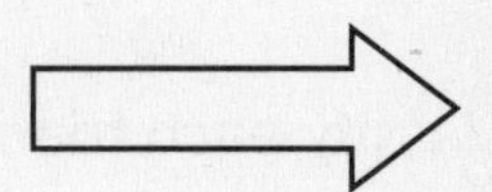

____

9. Choose the figure below that does not belong. Explain.

____

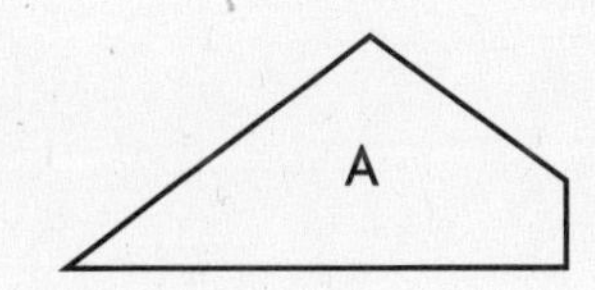

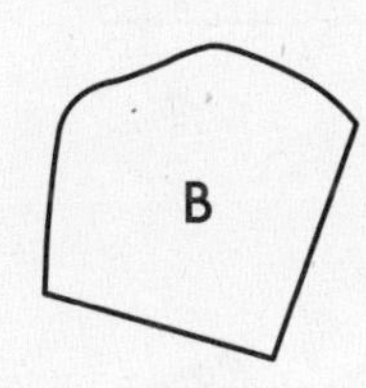

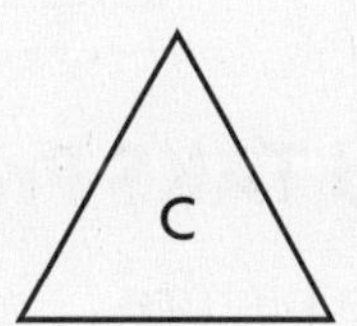

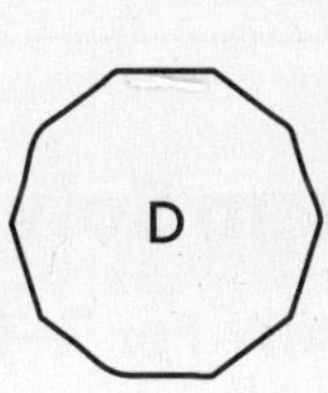

Problem Solving and TAKS Prep

USE DATA For 10–11, use the pattern at the right.

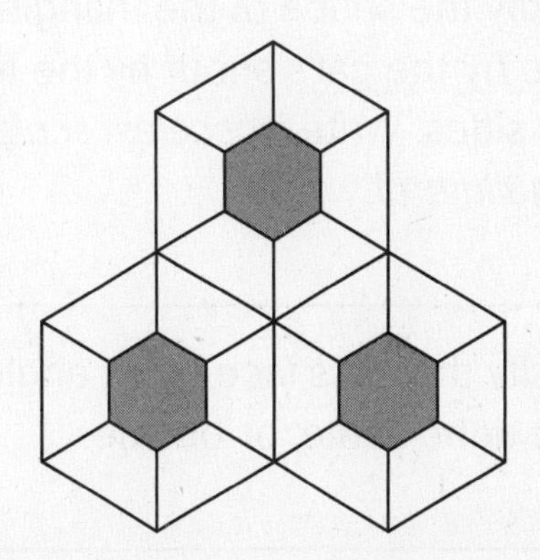

10. What is the name of the shaded polygon?

____

11. What other polygon do you see?

____

12. Which of the polygons have more than seven sides?

A triangle

B pentagon

C octagon

D hexagon

13. A yield sign has three sides of equal length and three angles of equal measure. What is a good description for this figure?

____

Name ______________________________ Lesson 10.5

Classify Triangles

Classify each triangle. Write *isosceles*, *scalene*, or *equilateral*. Then write *right*, *acute*, or *obtuse*.

1.

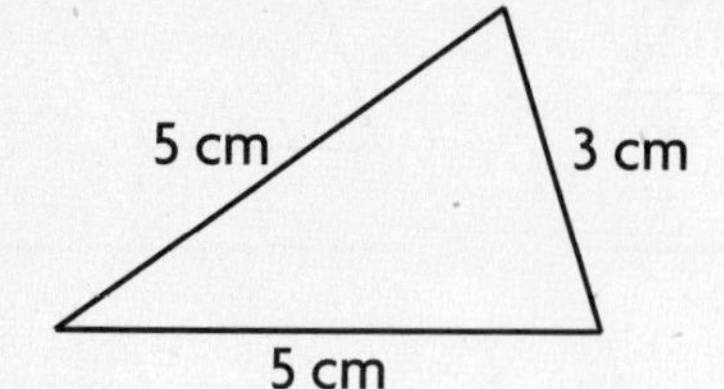

____

2.

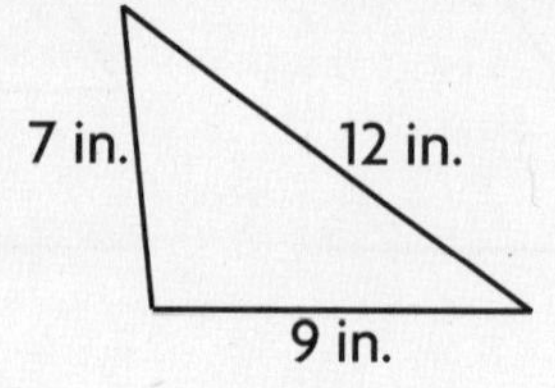

____

3.

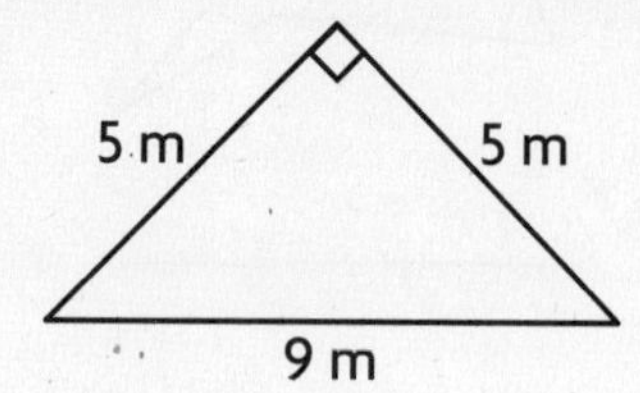

____

Classify each triangle by the lengths of its sides.

4.

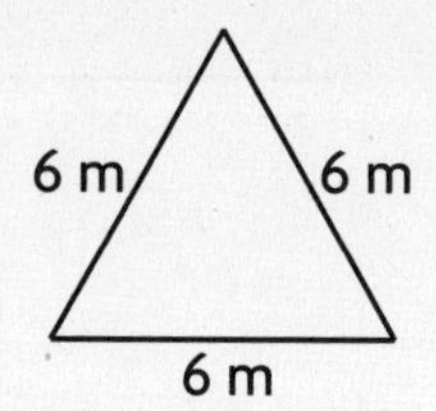

____

5.

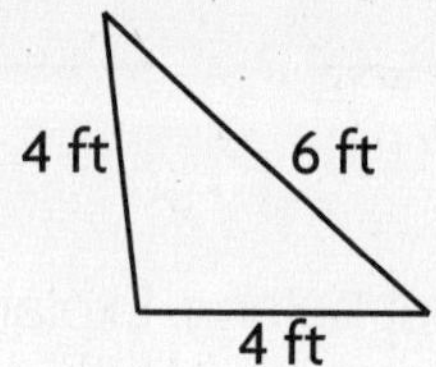

____

6.

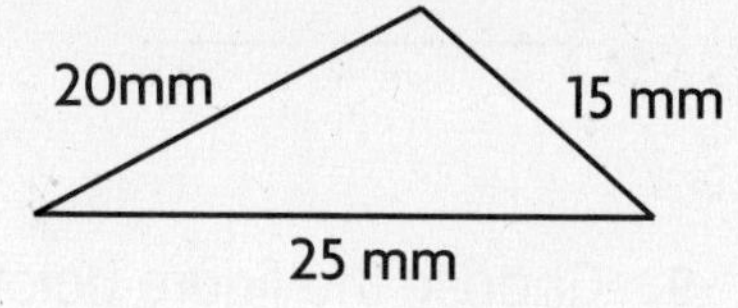

____

Problem Solving and TAKS Prep

USE DATA For 7–8, use the picture.

7. Classify the shape of the triangle made by the cat's snout by the length of its sides. Write *isosceles*, *scalene*, or *equilateral*.

____

8. Classify the cat's face by its angles. Write right, acute, or obtuse.

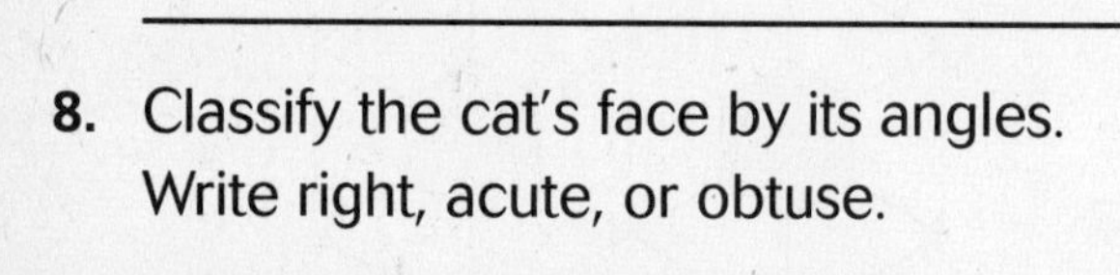

____

9. What kind of triangle has 2 equal sides?

A acute

B equilateral

C isosceles

D scalene

10. What kind of triangle has no equal sides?

F acute

G equilateral

H isosceles

J scalene

Name____ **Lesson 10.6**

Classify Quadrilaterals

Classify each figure in as many of the following ways as possible. Write ***quadrilateral, parallelogram, rhombus, rectangle, square,*** **or** ***trapezoid.***

1. ______________ ______________ ______________

2. ______________ ______________ ______________

3. ______________ ______________ ______________

4. ______________ ______________ ______________

Draw and label an example of each quadrilateral.

5. It has two pairs of parallel sides and opposite sides equal.

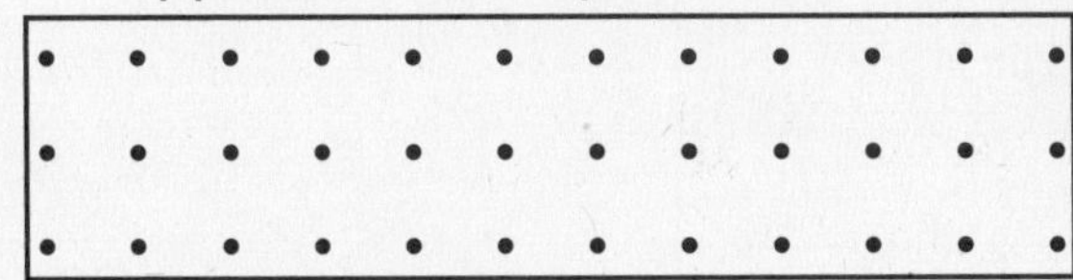

6. It has 4 equal sides with 4 right triangles.

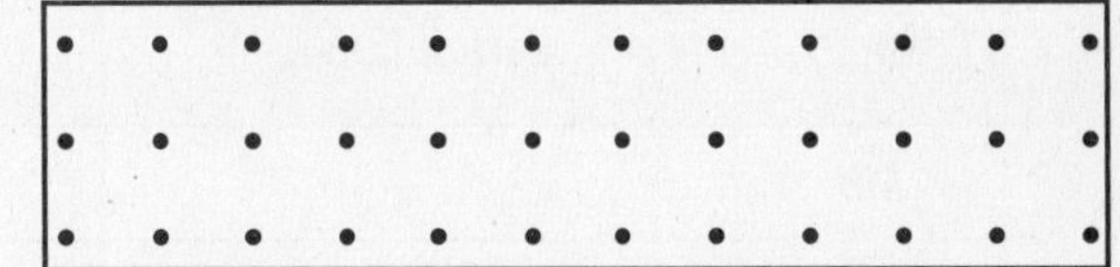

7. It has 4 equal sides with 2 pairs of parallel sides.

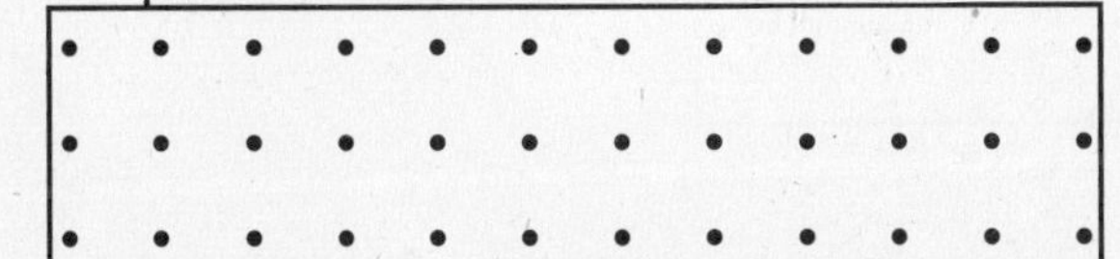

8. It has no pairs of parallel sides.

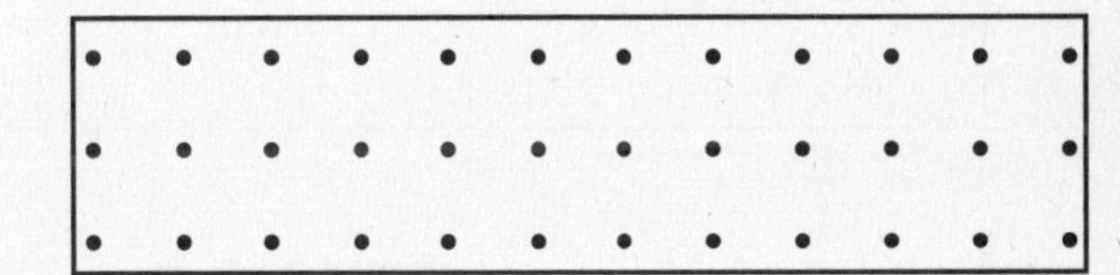

Problem Solving and TAKS Prep

USE DATA For 9–10, use the picture.

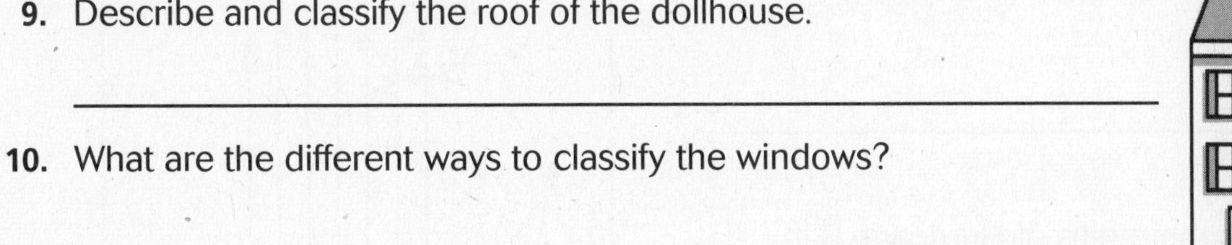

9. Describe and classify the roof of the dollhouse.

____

10. What are the different ways to classify the windows?

____

11. Which is the best description of the figures shown below?

A parallelograms

B quadrilaterals

C rectangles

D trapezoids

12. Which is the best description of the figures?

F parallelograms

G quadrilaterals

H rectangles

J trapezoids

Name________________________________ Lesson 10.7

Circles

In the box at the right, construct circle *M* with a radius of 2 cm. Label each of the following.

1. radius: $\overline{MB}$
2. diameter: $\overline{CD}$
3. radius: $\overline{ME}$
4. diameter: $\overline{KL}$

For 5–8, use the circle you drew and a centimeter ruler to complete the table.

	Name	Part of Circle	Length in cm
5.	$\overline{MB}$		
6.	$\overline{CD}$		
7.	$\overline{ME}$		
8.	$\overline{KL}$		

Problem Solving and TAKS Prep

USE DATA For 9–10, use the diagram.

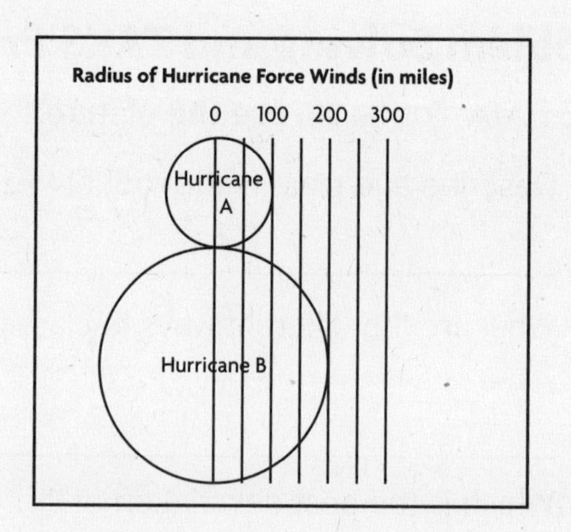

9. What is the diameter of Hurricane A in miles?

____

10. What is the radius of Hurricane B in miles?

____

11. What is the length of the diameter of a circle with a radius of 6 cm?

A 3 cm C 9 cm
B 6 cm D 12 cm

12. What is the length of the radius of a circle with a diameter of 16 ft?

F 32 ft H 4 ft
G 8 ft J 16 ft

Name________________________________ Lesson 10.8

Problem Solving Workshop Strategy: Use Logical Reasoning

Problem Solving Strategy Practice

For 1–3, use the figures at the right.

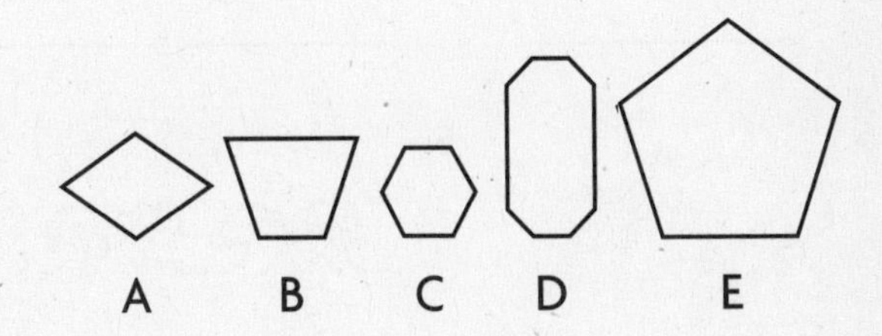

1. Lenny's parents put a play area in their backyard. All the sides of the play area are of equal length and none of the angles are acute or square. Identify the figure shown that appears to be like Lenny's play area.

   ____

2. Cyd is designing a garden that has no parallel sides and all obtuse angles. Identify the figure shown that appears to be like Cyd's design.

   ____

3. The shape of Holly's backyard has two parallel sides and two acute angles. Identify the figure shown that appears to be like Holly's backyard.

   ____

Mixed Strategy Practice

4. Wila and her two brothers each have the amount of money shown below. How much money does each person have?

   ____

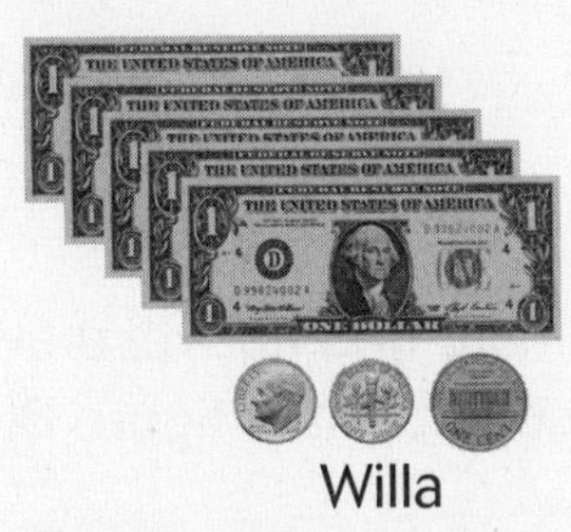

Willa

Bob

Jon

5. After Della tossed coins into a pool, James dove in to pick up the quarter. Then Della dove in to pick up her remaining 30 cents. How much money did Della toss into the pool?______________________

6. Han's backyard was shaped like a square with all right angles. Classify the design in as many ways as possible.

   ____

Name____ **Lesson 11.1**

Congruent Figures

Tell whether the two figures appear to be *congruent* or *not congruent*.

1. 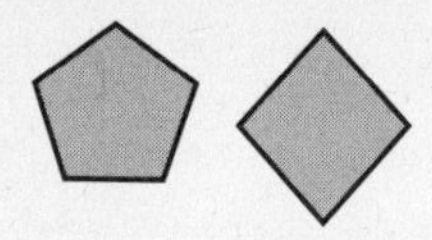

____

2. 

____

3. 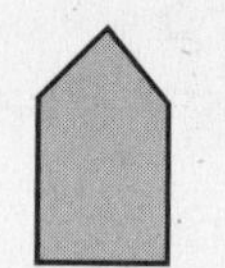

____

4. 

____

5. 

____

6. 

____

7. 

____

8. 

____

For 9–11, use the coordinate grid. Write *true* or *false*.

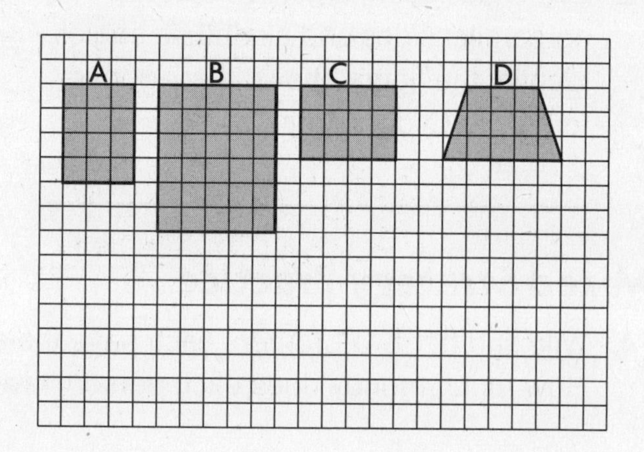

9. Which pair of rectangles appear to be congruent?

____

10. Which rectangle is not congruent to one of the others?

____

11. Use the dot paper to draw a figure that is congruent to figure B.

Problem Solving and TAKS Prep

USE DATA For 12–13, use the grid above.

12. Draw a diagonal line through figure B to make two triangles. Are the triangles congruent? ______________

13. On the coordinate grid above, draw a figure that is congruent to figure D.

____

14. What word or words best describes the figures?

A congruent

B pentagons

C not congruent

D none of the above

15. Which figures appear to be congruent?

F  

G  

H 

J 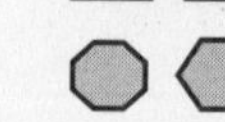

Name________________________________ **Lesson 11.2**

Translations

Tell if only a translation was used to move the figure. Write *yes* or *no*.

1.

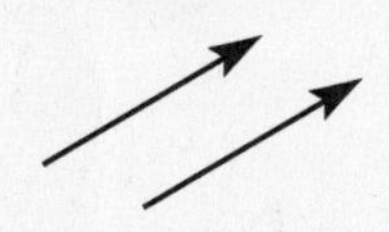

____

2.

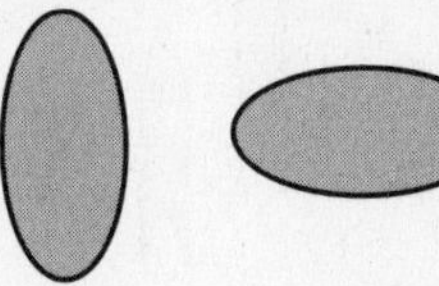

____

3.

____

4.

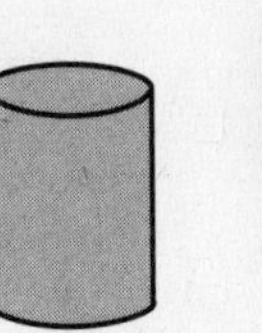

____

For 5–7, use the figures at the right. Write all the letters which make the statement true.

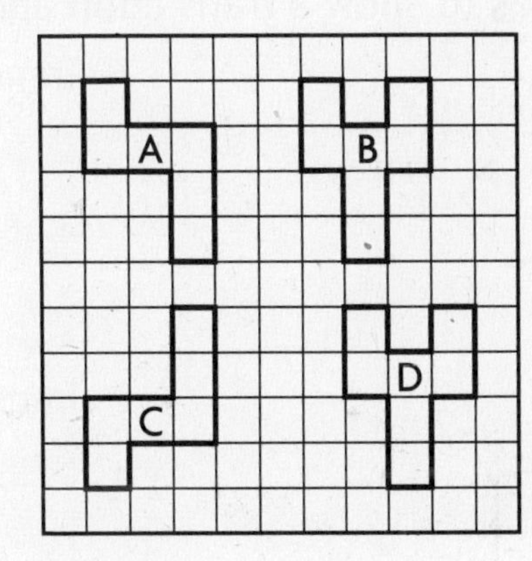

5. One figure is a translation of another figure.

____

6. The figures are congruent.

____

7. The figures are not congruent.

____

Draw a figure to show a translation of each figure.

8.

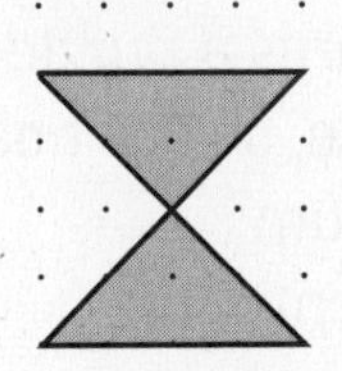

9.

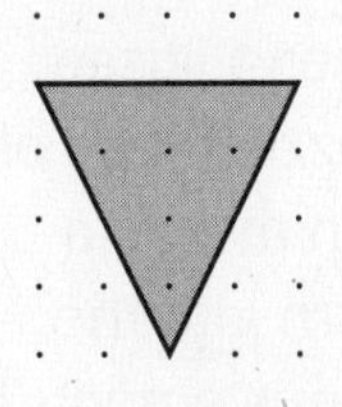

10.

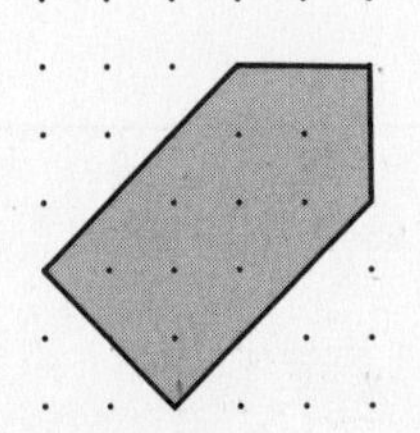

11.

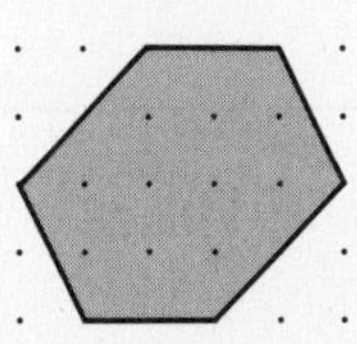

Name____

Rotations

Tell how each figure was moved. Write *translation* or *rotation*.

1. 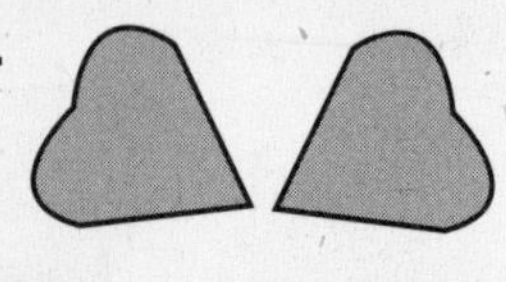

____

2. 

____

3. 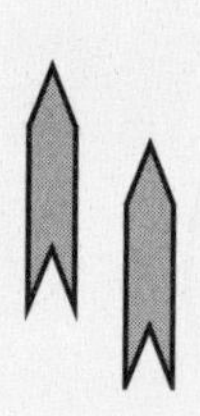

____

4. 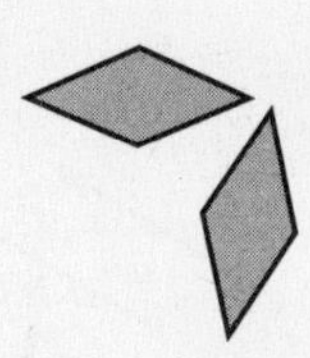

____

5. 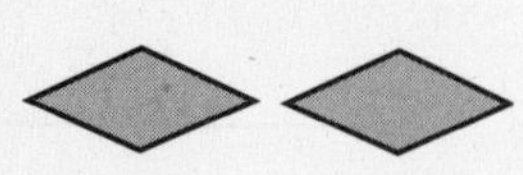

____

6. 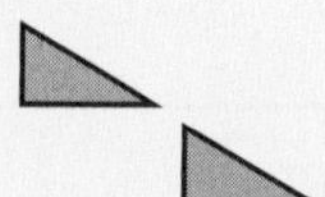

____

7. 

____

8. 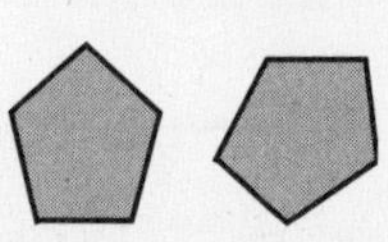

____

Draw figures to show a translation and then a rotation of the original.

9. 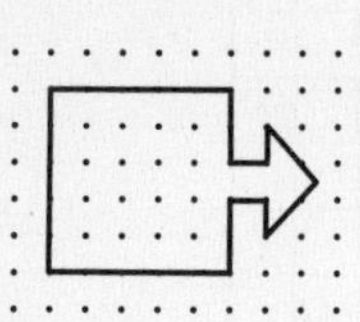

Translation: 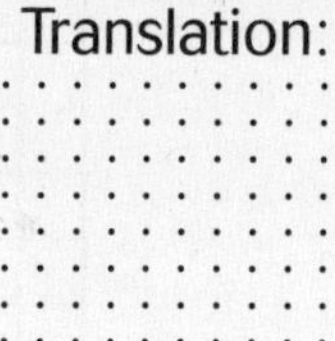

Rotation: 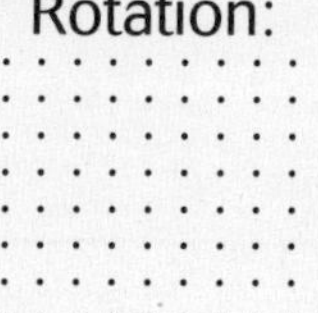

10. 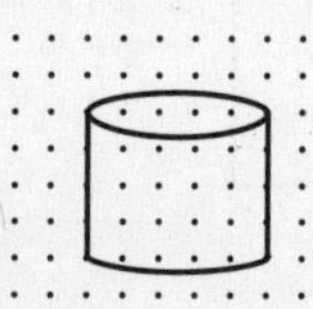

Translation:

Rotation: 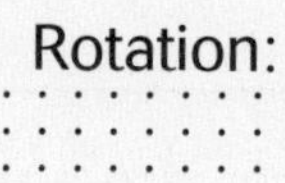

Problem Solving and TAKS Prep

11. Jules is putting together a jigsaw puzzle. He slides a puzzle piece along a straight line and then makes a counterclockwise $\frac{3}{4}$ turn into the puzzle. Identify these movements?

____

____

12. To help Jules, Julie turns a puzzle piece $\frac{1}{4}$ turn clockwise, slides it back, and then makes a $\frac{1}{2}$ turn counterclockwise. Identify the movements.

____

____

13. Which pair of figures are **not** congruent?

A
B

C
D

14. Which pair of figures is congruent?

F
G

H
J

Name____________________________________ **Lesson 11.4**

Reflections

Tell how each figure was moved. Write *translation*, *rotation*, or *reflection*.

1. 

____

2. 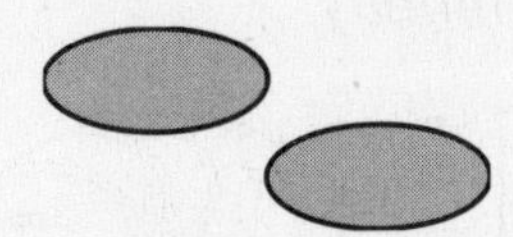

____

3. 

____

4. 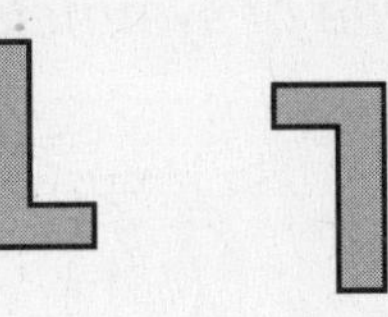

____

Use the figures below.
Write the letter of the figure that shows how it will look after each move.

5. 

Translation _____ Reflection _____ Rotation _____

a. 

b. 

c. 

6.

Translation _____ Reflection _____ Rotation _____

a.

b. 

c. 

Problem Solving and TAKS Prep

7. Use the dot paper. Draw a translation, rotation, and reflection of the first 3 letters in your name. Do any letters look the same when you translate, rotate, or reflect?

____

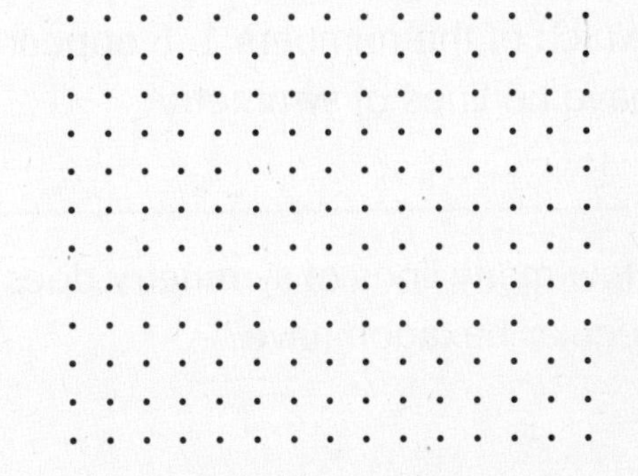

8. Make a pattern with at least two different moves using one of the letters in your name.

____

9. Which of the following shows a reflection?

A **M** **M**

B **W** **M**

C

D

10. Which of the following shows a reflection followed by a rotation?

F **G** **H** **J**

Name____ **Lesson 11.5**

Symmetry

Tell whether the figure has *no lines of symmetry, 1 line of symmetry, or more than 1 line of symmetry.*

1. 
____

2. 
____

3. 
____

4. 
____

Draw the line or lines of symmetry.

5. 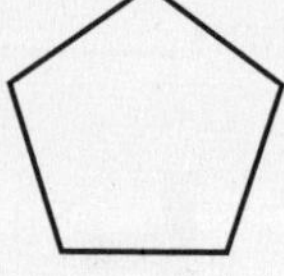

6. 

7. 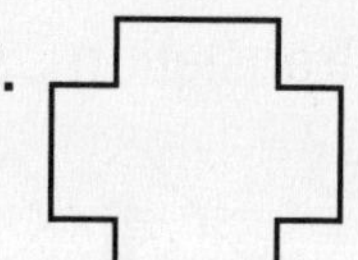

8. 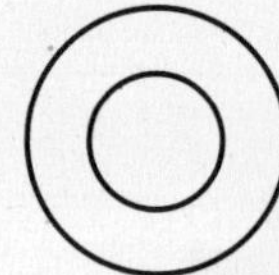

Complete each design to show symmetry.

9. 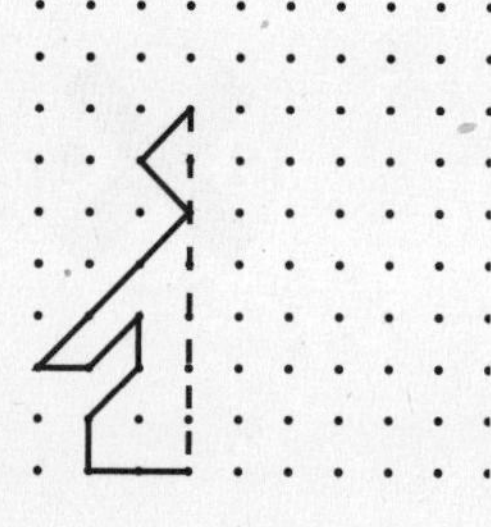

10. 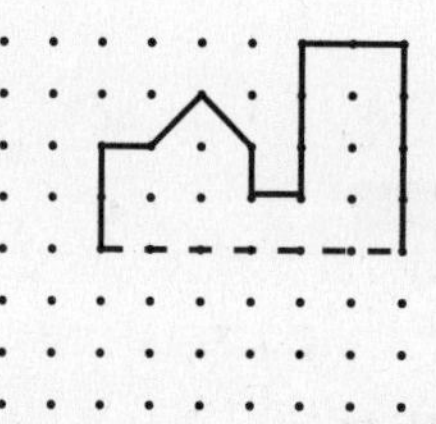

11. 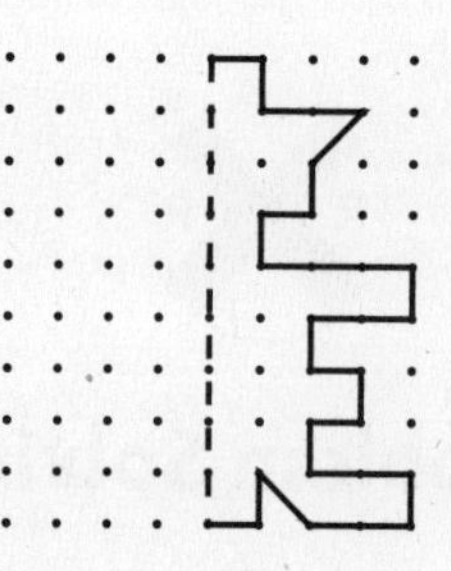

Problem Solving and Test Prep

12. Which of the numbers 1–8 appear to have no lines of symmetry?

____

13. Which of the numbers 1–8 appear to have more than one line of symmetry?

____

14. How many lines of symmetry does a regular hexagon have?

A 1
B 5
C 6
D 10

15. What letters and numbers in the license plate below appear to have lines of symmetry?

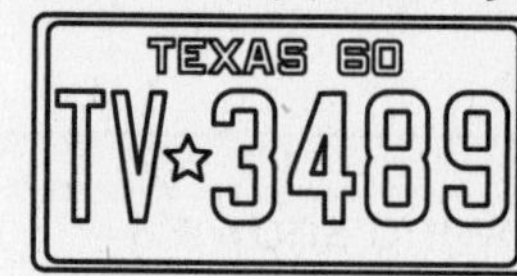

____

Name____ Lesson 11.6

Problem Solving Workshop Strategy: Act It Out

Problem Solving Strategy Practice

Act it out to solve.

1. Don made the two model airplanes shown at the right using different pattern blocks. Does each airplane have symmetry? What kind of symmetry?

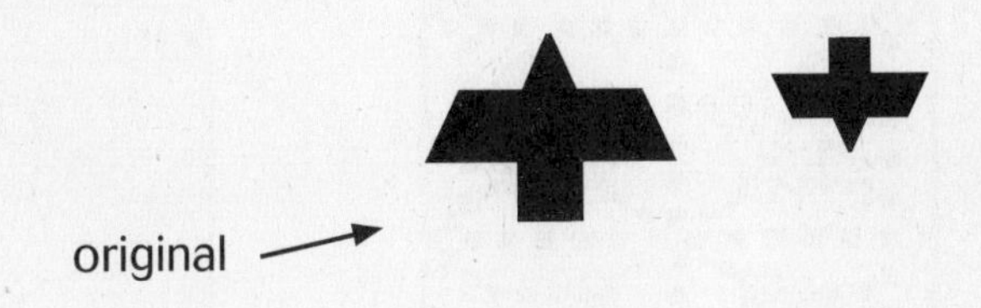

____

2. Jen, Bea, Sara, and Jon are in a math contest. Before it starts, each contestant gives one score card to each of the other contestants. How many scorecards are there in all?

____

3. Here is the design Craig is making.

What will the 20th item be in the design?

____

Mixed Strategy Practice

4. Starting at a base, Don flies his model airplane 50 feet south, 20 feet east, 10 feet north, and 20 feet west. How far is Don's plane from a flagpole that is 15 feet north of the base?

____

5. Fred uses 3 red, 2 blue, and 4 green squares to make one design for the border in his room. If he uses 108 squares in all, how many of each color will Fred use?

____

6. Lara spent $44 on supplies to make a curtain. She bought fabric for $25, backing for $8, and hem tape for $8, and thread. How much did the thread cost?

____

7. Dianna is making a list of the 120 seashells in her collection. What are some of the ways she might organize her list if she has 7 different colors, the shells came from 12 different beaches, and she has 3 display shelves?

____

____

____

____

Name________________________________

Tessellations

Trace and cut out several of each figure. Tell whether the figure will tessellate. Write *yes* or *no*.

1. ____________

2. ____________

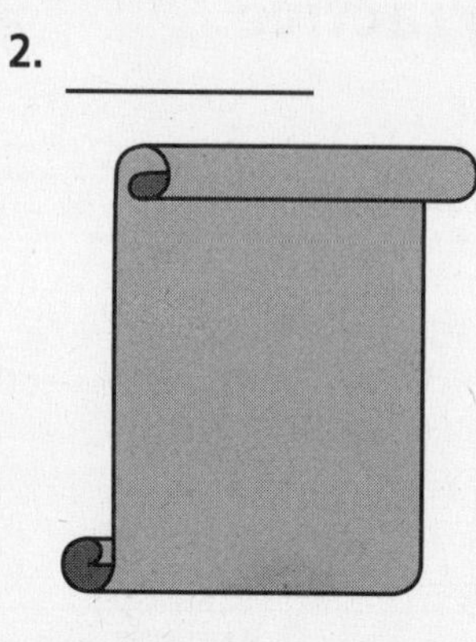

3. ____________

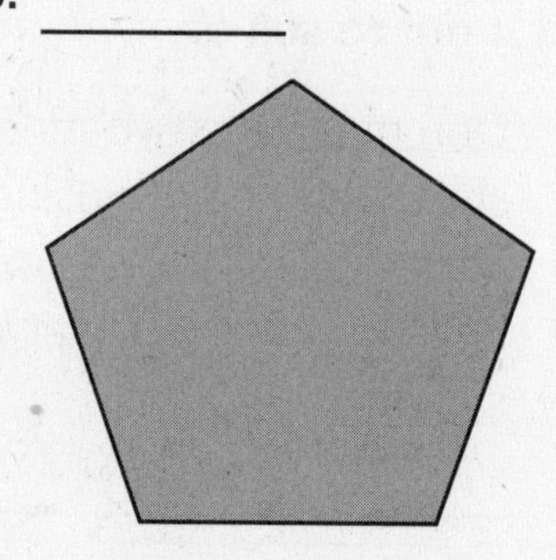

4. ____________

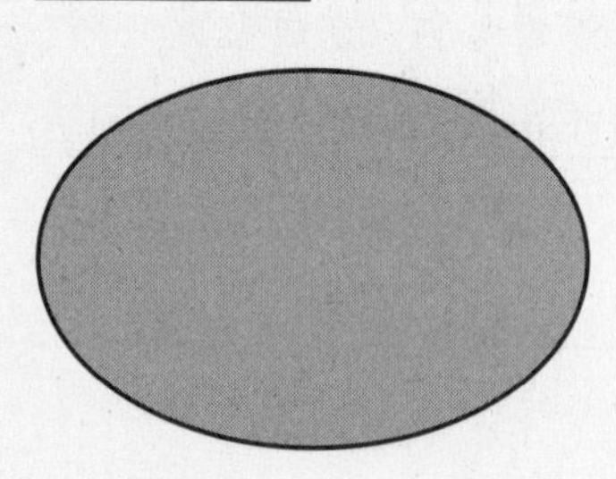

5. ____________

6. ____________

7. ____________

8. ____________

9. ____________

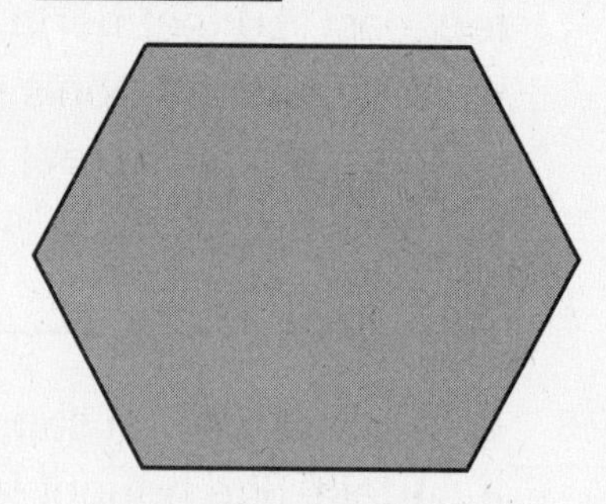

10. ____________

11. ____________

12. ____________

Name________________________________

Geometric Patterns

Find a possible pattern. Then draw the next two figures.

1.    ___ ___

2.  ___ ___

3.  ___ ___

4. ( )( )( )( )( ) ___ ___

Find a possible pattern. Then draw the missing figure.

5. ● ●● ●●● ●● ● ____ ●●●

6.   ___ 

7.  ___   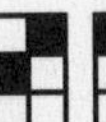 

Problem Solving and TAKS Prep

USE DATA For 8–9, use the quilt.

8. Does the rule for the pattern appear to include colors or shading? Explain.

9. If you remove the border and add a row at the bottom, will that row start with a block or a triangle?

10. In Exercise 6, what will be the tenth figure in the pattern?

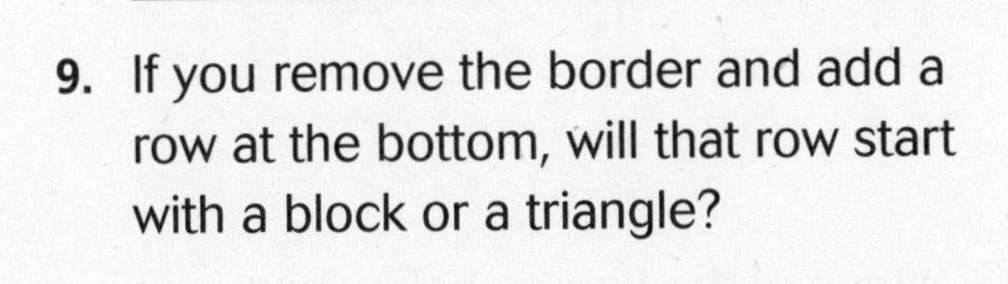

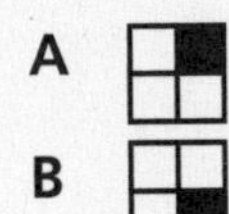

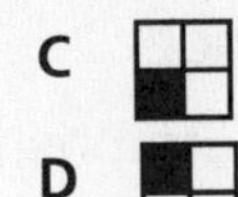

11. In Exercise 2, if the arrows continue to rotate, what will be the fifteenth figure in the pattern?

F ⇧ H ⇩

G ⇦ J ⇨

Name

Faces, Edges, and Vertices

Name a solid figure that is described.

1. 2 circular bases

2. 6 square faces

3. 1 rectangular and 4 triangular faces

4. 1 circular base

Which solid figure do you see in each?

5. 

6. 

7. 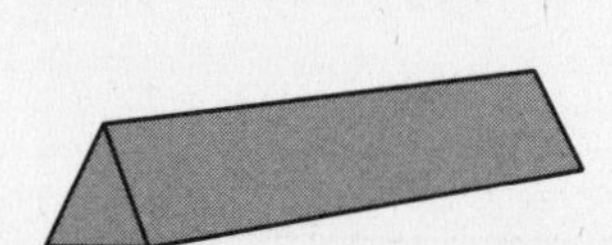

8. 

Problem Solving and TAKS Prep

For 9–11, use the rectangular prism.

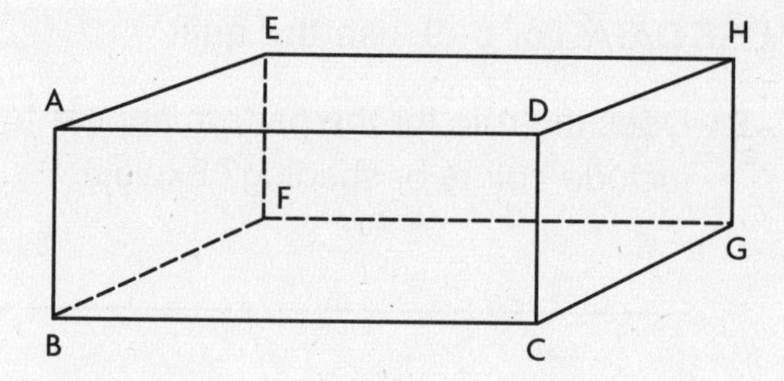

9. Name a pair of parallel line segments.

10. Name a pair of perpendicular line segments.

11. Which solid figure has more edges, a rectangular prism or a triangular prism? How many more?

12. What is the relationship between the number of faces and the number of edges of a triangular pyramid?

13. Which of the figures below has one face?
   - **A** cone
   - **B** sphere
   - **C** cylinder
   - **D** square prism

Name ______________________

Lesson 12.2

Draw Solid Figures

For 1–5, use the table.

1. What solid figure does a textbook look like?

____

Item	Vertices	Edges	Faces
Textbook	8	12	6 rectangles
pencil	12	18	6 rectangles 2 hexagons

2. Use dot paper to draw a textbook.

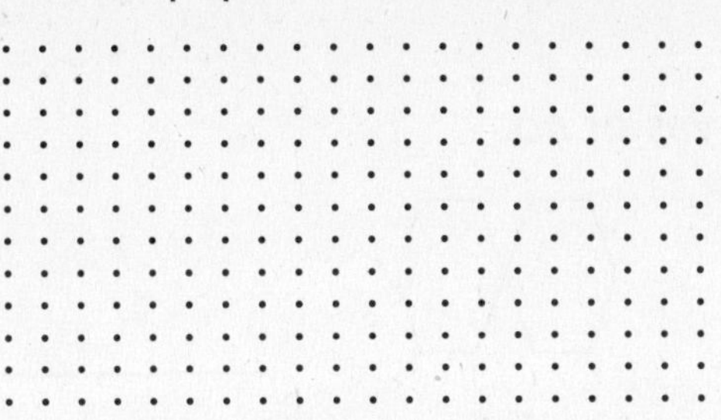

3. Use dot paper to draw a pencil.

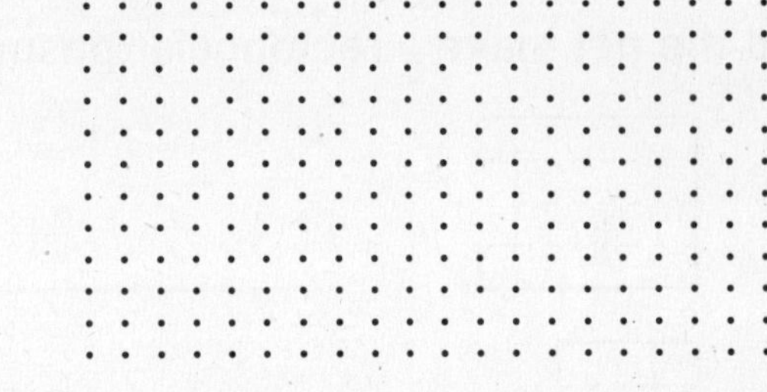

Problem Solving and TAKS Prep

For 4–5, use the dot paper at the right to draw each figure. Label the vertices. Identify any parallel and perpendicular line segments you see in each figure.

4. **a.** a square with sides 2 units long

b. a square pyramid rising from the square

____

____

5. **a.** a rectangular prism with one edge 3 units long

b. a cube with sides 2 units long

____

____

6. How many line segments do you need to draw a cube?

A 12 **C** 6

B 8 **D** 4

7. How many line segments do you need to draw a pencil?

F 12 **H** 16

G 14 **J** 18

Name________________________________

Patterns for Solid Figures

Draw a net that can be cut to make a model of each solid figure.

1.

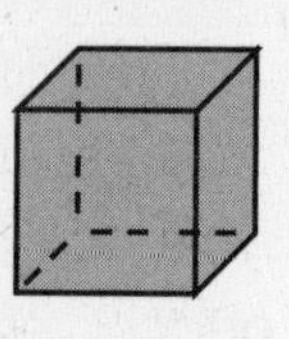

2.

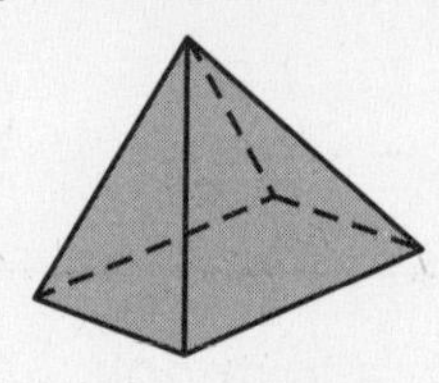

Would the net make a rectangular prism? Write *yes* or *no*.

3.

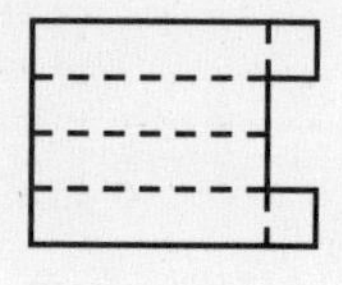

____

4.

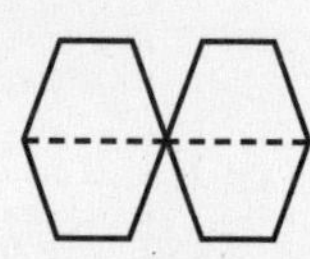

____

5.

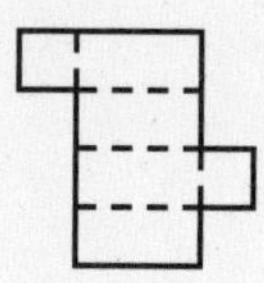

____

6.

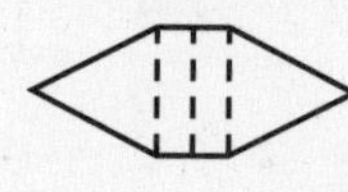

____

For 7–8, use the nets at the right.

7. Do nets B and C make figures with the same number of sides?

____

8. Do nets A and C make figures with the same number of edges? Explain.

____

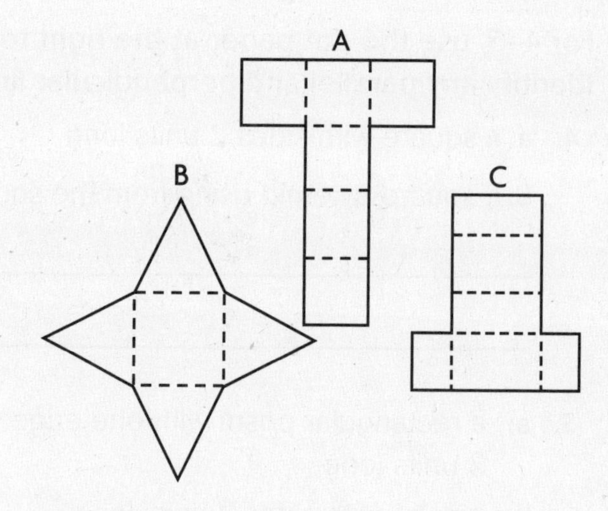

Problem Solving and TAKS Prep

9. How would you change the figure in Exercise 3 to make a solid figure?

____

10. Can the net in Exercise 6 make a solid figure?

____

11. What figure can you make when net A is folded on the dotted lines without overlapping?

____

12. What figure can you make when net B is folded on the dotted lines without overlapping?

____

Name________________________________ **Lesson 12.4**

Problem Solving Workshop Strategy: Make a Model

Problem Solving Strategy Practice

Make a model to solve.

1. Paula has 36 cubes to build a wall that is 1, 2, and 3 cubes high and then repeats the pattern. How many cubes long can Paula make the wall?

____

2. What if Paula used a repeating pattern of 1, 3, and 5 blocks high? How many blocks would Paula need to build a wall 9 blocks long?

____

3. John has 66 cubes. He gives 21 to Mark and then builds a staircase beginning with 1 cube, then 2, and so on. How tall will John's staircase be?

____

4. How many cubes would John need to build the next step of his staircase?

____

Mixed Strategy Practice

5. Sandra and Jan have a total of 88 cubes, half of which are blue. Jan uses 34 to make a wall and Sandra uses 25 to make a building. What is the least number of blue cubes they could use?

____

6. Mrs. Lutie left home and went to the bank. Then she drove 18 miles to the dentist, 9 miles for groceries, 8 miles to pick up her kids, and 3 miles back home. If Mrs. Lutie drove a total of 45 miles, how far was it from home to the bank?

____

7. **Pose a Problem** Change the numbers in Exercise 6. Make a new problem about Mrs. Lutie's errands.

____

____

____

8. How many ways can you arrange 12 cubes in more than one row? Name the ways.

____

____

____

Name____

Divide with Remainders

Use counters to find the quotient and remainder.

1. $27 \div 5 =$ ■ r ■ **2.** $34 \div 8 =$ ■ r ■ **3.** $18 \div 4 =$ ■ r ■

4. $57 \div 7 =$ ■ r ■ **5.** $41 \div 6 =$ ■ r ■ **6.** $53 \div 9 =$ ■ r ■

7. ■ r ■ $3\overline{)26}$ **8.** ■ r ■ $7\overline{)64}$ **9.** ■ r ■ $9\overline{)87}$

Divide. You may wish to use counters or draw a picture to help.

10 ■ r ■ $3\overline{)26}$ **11.** ■ r ■ $4\overline{)34}$ **12.** ■ r ■ $6\overline{)50}$

13. $75 \div 9 =$ ■ r ■ **14.** $54 \div 8 =$ ■ r ■. **15.** $60 \div 7 =$ ■ r ■

16. ■ r ■ $3\overline{)17}$ **17.** ■ r ■ $5\overline{)44}$ **18.** ■ r ■ $7\overline{)33}$

19. $19 \div 2 =$ ■ r ■ **20.** $50 \div 8 =$ ■ r ■ **21.** $41 \div 9 =$ ■ r ■

Model 2-Digit by 1-Digit Division

Use base-ten blocks to find the quotient and remainder.

1. $37 \div 2 =$ ■r■ **2.** $53 \div 5 =$ ■r■ **2.** $92 \div 7 =$ ■r■ **4.** $54 \div 4 =$ ■r■

5. $56 \div 3 =$ ■r■ **6.** $89 \div 9 =$ ■r■ **7.** $78 \div 6 =$ ■r■ **8.** $92 \div 8 =$ ■r■

9. $4\overline{)65}$ ■ r ■ **10.** $7\overline{)79}$ ■ r ■ **11.** $6\overline{)89}$ ■ r ■ **12.** $4\overline{)87}$ ■ r ■

Divide. You may wish to use base-ten blocks.

13. $3\overline{)77}$ ■ r ■ **14.** $2\overline{)67}$ ■ r ■ **15.** $4\overline{)64}$ ■ r ■ **16.** $5\overline{)67}$ ■ r ■

17. $37 \div 2 =$ ■r■ **18.** $98 \div 4 =$ ■r■ **19.** $91 \div 6 =$ ■r■ **20.** $72 \div 7 =$ ■r■

21. $8\overline{)93}$ ■ r ■ **22.** $6\overline{)57}$ ■ r ■ **23.** $4\overline{)77}$ ■ r ■ **24.** $9\overline{)59}$ ■ r ■

Name____________________

Lesson 13.3

Record 2-Digit by 1-Digit Division

Choose a method. Then divide and record.

1. $4\overline{)93}$ ▢▢ r ▢

2. $7\overline{)75}$ ▢▢ r ▢

3. $5\overline{)97}$ ▢▢ r ▢

4. $49 \div 3 =$ ▢ r ▢

5. $61 \div 2 =$ ▢ r ▢

6. $95 \div 7 =$ ▢ r ▢

7. $9\overline{)87}$ ▢▢ r ▢

8. $6\overline{)87}$ ▢▢ r ▢

9. $8\overline{)99}$ ▢▢ r ▢

ALGEBRA Complete each table.

10.

Number of Cups	16	20	24	28	32
Number of Quarts	4	5	▢	▢	▢

11.

Number of Pints	64	72	80	88	96
Number of Gallons	8	9	▢	▢	▢

Problem Solving and TAKS Prep

12. Sixty-three students signed up for golf. The coach divided them into teams of 4. How many students were left over?

____

13. There are 6 runners on each relay team. If a total of 77 runners signed up, how many relay teams could there be?

____

14. Four students divided 85 base-ten blocks equally among them. How many base-ten blocks does each student receive?

A 20

B 21

C 22

D 24

15. Three students divided 85 base-ten rods equally among them. How many base-ten rods were left over?

F 4

G 3

H 2

J 1

Name ______________________

Problem Solving Workshop Strategy: Compare Strategies

Problem Solving Strategy Practice

Choose a strategy to solve the problems.

1. Fiona's dog is 4 times as long as Rod's dog. End by end, they are 60 inches long. How long is Fiona's dog?

____

2. Davey divided 112-ounces of rabbit food equally into 7 containers. How much did each container hold?

____

3. Dina walked from home 3 blocks west and 5 blocks north to get to the pet store. If now she walks 1 block east, 4 blocks north, and another 2 blocks east, how far is Dina from home?

____

4. Mel is collecting 7 cards each of his 13 favorite baseball players. He now has a total of 87 cards. How many additional cards does Mel need to make his set of 7 cards each complete?

____

Mixed Strategy Practice

USE DATA For 5 – 6, use the chart.

Dog Heights

Breed	Height
Bichon Frise	10 in.
Border Collie	20 in.
Chihuahua	8 in.
Irish Setter	27 in.
Labrador Retriever	24 in.
Shar-Pei	19 in.
Siberian Huskey	22 in.

5. Together the height of Dan's 3 dogs is 38 inches. What breeds are they?

____

6. Order the dogs in the table from shortest to tallest.

____

____

7. Altogether, Haille's dog statue collection weighs 20 pounds. One statue weighs 8 pounds and the rest weigh half as much. How many dog statues does Haille have?

____

8. **Pose a Problem** Use the information from Exercise 5 to write a new problem that asks to explain the answer. ______________________

____

Name______________________________ **Lesson 13.5**

Mental Math: Division Patterns

Use mental math to complete the pattern.

1. $72 \div 8 = 9$
$720 \div 8 =$ ______
$7{,}200 \div 8 =$ ______
$72{,}000 \div 8 =$ ______

2. $42 \div 7 =$ ______
______ $\div 7 = 60$
$4{,}200 \div 7 =$ ______
$42{,}000 \div 7 =$ ______

3. ______ $\div 6 = 4$
$240 \div 6 =$ ______
______ $\div 6 = 400$
$24{,}000 \div 6 =$ ______

4. $30 \div 3 =$ ______
______ $\div 3 = 100$
$3{,}000 \div 3 =$ ______
______ $\div 3 = 10{,}000$

5. ______ $\div 5 = 8$
$400 \div 5 =$ ______
______ $\div 5 = 800$
$40{,}000 \div 5 =$ ______

6. $28 \div 4 =$ ______
______ $\div 4 = 70$
$2{,}800 \div 4 =$ ______
______ $\div 4 = 7{,}000$

Use mental math and patterns to find the quotient.

7. $1{,}600 \div 4 =$ ______

8. $28{,}000 \div 7 =$ ______

9. $50 \div 5 =$ ______

10. $900 \div 3 =$ ______

11. $32{,}000 \div 4 =$ ______

12. $2{,}000 \div 5 =$ ______

13. $600 \div 2 =$ ______

14. $3{,}500 \div 7 =$ ______

Problem Solving and TAKS Prep

15. Maria has 4,500 stamps in her collection. She puts an equal amount of stamps into 9 books. How many stamps will be in each book?

____

16. Tex wants to put 640 stickers in his sticker book. If there are 8 stickers to a page, how many pages will Tex fill?

____

17. The theme park sells tickets for $4 each. It collects $2,000 in one day. How many tickets does the park sell in one day?

A 50
B 500
C 5,000
D 50,000

18. Dee collected $60 for selling tickets. If she sold 5 tickets, how much did each ticket cost?

F $12
G $24
H $30
J $30

Name ____________________

Mental Math: Estimate Quotients

Estimate the quotient.

1. $392 \div 4$ ________
2. $489 \div 6$ ________
3. $536 \div 9$ ________
4. $802 \div 8$ ________
5. $632 \div 7$ ________
6. $32{,}488 \div 4$ ________
7. $3{,}456 \div 5$ ________
8. $7{,}820 \div 8$ ________

Estimate to compare. Write <, >, or = for each ●.

9. $272 \div 3$ ● $460 \div 5$
10. $332 \div 6$ ● $412 \div 5$
11. $527 \div 6$ ● $249 \div 3$
12. $138 \div 2$ ● $544 \div 9$
13. $478 \div 7$ ● $223 \div 3$
14. $3{,}112 \div 8$ ● $1{,}661 \div 8$

Problem Solving and TAKS Prep

USE DATA For 15–16, use the table.

15. Which beats faster, a dog's heart in 5 minutes or a mouse's heart in 1 minute?

____

16. Which beats slower in 1 minute: a human's heart or a horse's heart?

____

Resting Heartbeats of Select Mammals

Mammal	Rate per 5 minutes
Human	375
Horse	240
Dog	475
Mouse	2,490

17. A common loon beats its wings about 1,250 times in 5 minutes. What is the best estimate of the number of times its heart beats in one minute?

A 20
B 40
C 250
D 400

18. Nine equal-length Arizona Black Rattlesnakes laid in a row measure 378 inches. What is the best estimate of the length of 1 rattlesnake?

F 20
G 40
H 200
J 400

Name ____________________

Place the First Digit

Tell where to place the first digit. Then divide.

1. $4\overline{)511}$

2. $7\overline{)621}$

3. $2\overline{)124}$

4. $3\overline{)423}$

5. $136 \div 2$

6. $215 \div 5$

7. $468 \div 6$

8. $357 \div 8$

Divide.

9. $3\overline{)166}$

10. $9\overline{)785}$

11. $4\overline{)334}$

12. $6\overline{)577}$

13. $116 \div 2$

14. $425 \div 5$

15. $627 \div 7$

16. $436 \div 8$

Problem Solving and TAKS Prep

17. Petra picked 135 petals from the flowers of sweet pea plants. Each flower has 5 petals. How many flowers did Petra pull petals from?

18. Todd wants to plant some thyme equally in 8 areas in his garden. If he has 264 plants, how many thyme plants can Todd put in each area?

19. In which place is the first digit in the quotient $118 \div 4$?
   - A ones
   - B tens
   - C hundreds
   - D thousands

20. In which place is the first digit in the quotient $1{,}022 \div 5$?
   - F ones
   - G tens
   - H hundreds
   - J thousands

Name____ Lesson 14.1

Problem Solving Workshop Skill: Interpret the Remainder

Problem Solving Skill Practice

Solve. Write *a*, *b*, or *c* to explain how to interpret the remainder.

a. **Quotient stays the same. Drop the remainder.**

b. **Increase the quotient by 1.**

c. **Use the remainder as the answer.**

1. The crafts teacher gave 8 campers a total of 55 beads to make necklaces. If he divided the beads equally among the campers, how many did each camper have?

   ____

2. In all, campers from each of 3 tents brought 89 logs for a bonfire. Two tents brought equal amounts but the third brought more. How much more?

   ____

3. Gene had 150 cups of water to divide equally among 9 campers. How many cups did he give each camper?

   ____

4. Camp leaders divided 52 cans of food equally among 9 campers. How many cans of food were left over?

   ____

Mixed Applications

5. Geena had 34 hot dogs. She gave 3 camp counselors 2 hot dogs each before dividing the rest equally among the 7 campers. How many hot dogs did she give each camper?

   ____

6. The morning of a hiking trip the temperature was 54°F. By mid afternoon, the temperature had risen to 93°F. How much warmer was the afternoon temperature?

   ____

7. **Pose a Problem** Exchange the known for unknown information in Exercise 5 to write a new problem.

   ____

   ____

8. Wynn bought these camping tools: a flashlight, an axe for $15, a lantern for $12, and a camp stool for $23. If he spent $57, how much did the flashlight cost?

   ____

Name________________________________

Lesson 14.2

Divide 3-Digit Numbers

Divide and check.

1. $147 \div 5 =$ ______ **2.** $357 \div 7 =$ ______ **3.** $575 \div 4 =$ ______

4. $6\overline{)844}$ **5.** $9\overline{)874}$ **6.** $8\overline{)766}$

ALGEBRA Find the missing digit.

7. $577 \div \blacksquare = 115$ r2 **8.** $\blacksquare 10 \div 2 = 405$ **9.** $734 \div 3 = 24\blacksquare$ r2 **10.** $572 \div 6 = \blacksquare 5$ r2

11. $9\overline{)593}$ = ■5 r8 **12.** $4\overline{)5\blacksquare 2}$ = 145 r5 **13.** $\blacksquare\overline{)572}$ = 71 r4 **14.** $7\overline{)488}$ = 69 r■

Problem Solving and TAKS Prep

15. In all, Alfred paid $18 for 12 bundles of asparagus at a local grocery store. If the bundles were in a buy-one-get-one-free sale, how much did each bundle cost before the sale?

____

16. Eva wants to divide 122 yards of yarn into 5-yard lengths to make potholders. How many potholders can Eva make? How many yards will be left over?

____

17. Ed divided 735 football cards among 8 friends. How many cards did each friend get?

A 98

B 91r7

C 99

D 99r3

18. Four cans of artichoke hearts are on sale for 12 dollars. How much does one can cost?

____

Name________________________________ **Lesson 14.3**

Zeros in Division

Write the number of digits in each quotient.

1. $366 \div 3$ ______

2. $5\overline{)374}$ ______

3. $635 \div 7$ ______

4. $4\overline{)923}$ ______

5. $672 \div 8$ ______

6. $5\overline{)811}$ ______

7. $9 \div 921$ ______

8. $6\overline{)597}$ ______

9. $816 \div 2$ ______

10. $7\overline{)177}$ ______

Divide and check.

11. $495 \div 5 =$ ______

12. $719 \div 6 =$ ______

13. $3\overline{)735}$

14. $4\overline{)897}$

15. $210 \div 4 = \blacksquare$

16. $103 \div \blacksquare = 14$ r5

17. $\blacksquare \div 5 = 61$

Problem Solving and TAKS Prep

18. Yoshi has a collection of 702 miniature cars that he displays on 6 shelves in his bookcase. If the cars are divided equally, how many are on each shelf?

______

19. In 5 days, scouts made a total of 865 trinkets for a fundraiser. If they made the same number each day, how many did they make in 1 day?

______

20. Greta has 594 flyers in stacks of 9 flyers each. How do you find the number of stacks Greta made? Explain.

______

______

21. Susan has 320 slices of banana bread. She wants to fill bags with 8 slices of banana bread each. How many bags will Susan fill?

______

______

Name________________________________ Lesson 14.4

Choose a Method

Divide. Write the method you used.

1. $2\overline{)643}$
2. $6\overline{)2,418}$
3. $4\overline{)6,458}$
4. $5\overline{)1,467}$
5. $3\overline{)2,483}$
6. $7\overline{)8,123}$
7. $8\overline{)7,467}$
8. $3\overline{)5,105}$
9. $7\overline{)6,111}$
10. $4\overline{)9,600}$

ALGEBRA Find the dividend.

11. ■ ÷ 3 = 178 ________________

12. ■ ÷ 4 = 733 ________________

13. ■ ÷ 7 = 410 ________________

14. ■ ÷ 9 = 245 r5 ________________

15. ■ ÷ 6 = 637 r1 ________________

16. ■ ÷ 8 = 801 r4 ________________

Problem Solving and TAKS Prep

17. Leona's team scored a total of 854 points in 7 days. Pilar's team scored a total of 750 points in 6 days. Which team scored more points per day?

____

18. Vicki has 789 seeds to put into packets. If she puts 9 seeds in each packet, how many packets will Vicki need?

____

19. Seth pledged a total of $3,336 over 6 months to a charity. How much will Seth donate each month?

A 210
B 333
C 336
D 556

20. Joe computed that he drove 1,890 miles a year roundtrip, to and from work. If his commute is 9 miles, how many days did he work?

A 210
B 333
C 336
D 556

Name______________________________ Lesson 14.5

Find the Average

Find the average.

1. 21, 18, 7, 28, 22, 24 ________

2. 4, 9, 12, 18, 27, 32 ________

3. 65, 41, 28, 37, 89, 70 ________

4. 49, 82, 100, 105, 124 ________

5. 188, 133, 127, 158, 164 ________

6. 111, 135, 67, 138, 199 ________

ALGEBRA Find the missing number.

7. 5, 1, 8, ■
The average is 4.

8. 6, 1, 8, ■
The average is 6.

9. 7, 14, 11, ■
The average is 9.

10. 8, 28, 17, 13, ■
The average is 16.

11. 10, 15, 31, 25, ■
The average is 20.

12. 23, 17, 35, 42, ■
The average is 26.

Problem Solving and TAKS Prep

USE DATA For 13–16, use the table.

13. Suppose Tim scored a 12 on his math test. How does his score compare with the average score on the math test?

____

14. Suppose the average math test score is 10. What would Tim's score be?

____

Class Test Scores			
Name	Spelling	Math	Reading
Jim	8	11	8
Bambi	12	10	10
Troy	15	10	13
Alice	9	9	11
Pan	6	14	13
Tim	?	?	?

15. Excluding Tim's score, what is the average spelling test score?

____

16. Excluding Tim's score, what is the average reading test score?

____

Name____

Lesson 15.1

Read and Write Fractions

Write a fraction for the shaded part. Write a fraction for the unshaded part.

1. 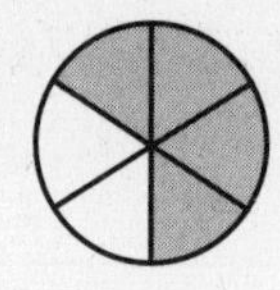

2. 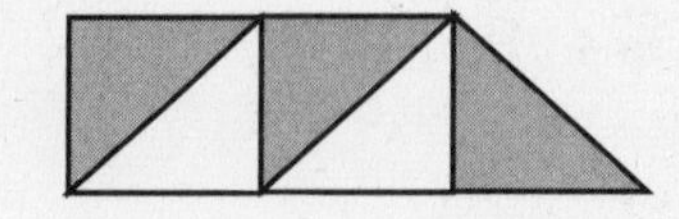

3. 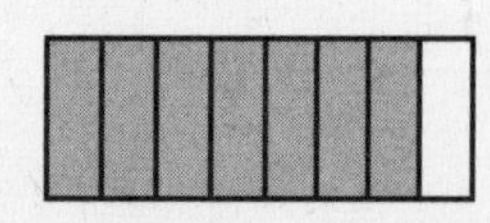

_______________ _______________ _______________

Draw a picture and shade part of it to show the fraction. Write a fraction for the unshaded part.

4. $\frac{5}{6}$

5. $\frac{4}{10}$

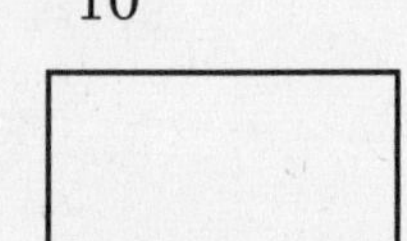

6. $\frac{3}{7}$

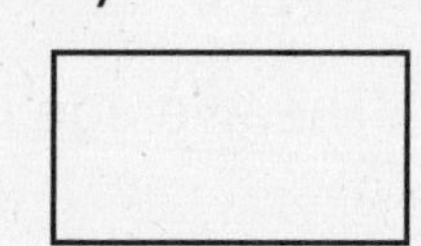

7. $\frac{3}{5}$

Write the fraction for each.

8. one-eighth

9. seven-tenths

10. four out of five

11. two divided by three

__________ __________ __________ __________

Problem Solving and TAKS Prep

12. Angela has 5 dollars to spend on lunch. She spends 1 dollar on a soda, 3 dollars on a hot dog, and 1 dollar on a bag of pretzels. What fraction of Angela's money does she spent on a hot dog?

____

13. There are 9 houses on Zach's block. 4 of them are red brick and the rest are gray brick. What fraction of the houses on Zach's block are gray brick?

____

14. Three friends cut a pizza into eight equal parts. The friends eat 3 pieces. What fraction of their pizza is left?

A $\frac{1}{8}$

B $\frac{3}{8}$

C $\frac{3}{5}$

D $\frac{5}{8}$

15. Melissa buys 3 apples, 4 pears, and 2 bananas from a fruit stand. What fraction of Melissa's fruit are pears?

F $\frac{3}{9}$

G $\frac{4}{9}$

H $\frac{2}{9}$

J $\frac{9}{9}$

Name ______________________

Model Equivalent Fractions

Write two equivalent fractions for each model.

1. 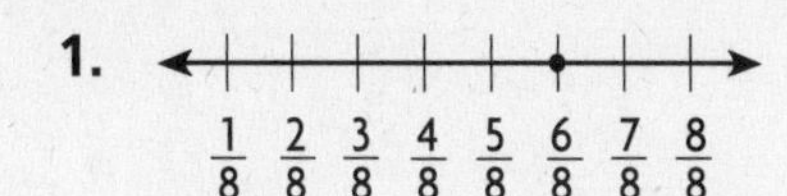

2. 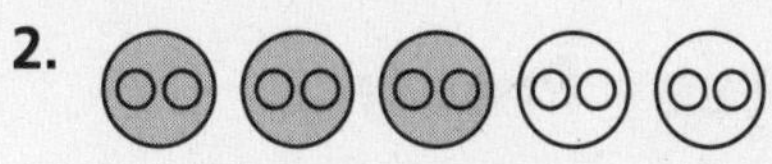

3. 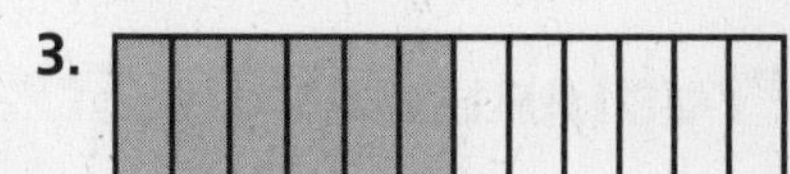

______________ ______________ ______________

Write two equivalent fractions for each. You may use a model or drawing.

4. $\frac{1}{5}$ ______________
5. $\frac{2}{3}$ ______________
6. $\frac{3}{12}$ ______________
7. $\frac{6}{8}$ ______________

Tell whether the fractions are equivalent. Write *yes* or *no*. You may use a model or drawing.

8. $\frac{2}{9}, \frac{4}{16}$ ________
9. $\frac{2}{6}, \frac{8}{24}$ ________
10. $\frac{1}{7}, \frac{2}{14}$ ________
11. $\frac{6}{12}, \frac{2}{3}$ ________

Tell whether the fraction is in simplest form. If not, write it in simplest form.

12. $\frac{12}{16}$ ______________
13. $\frac{5}{9}$ ______________
14. $\frac{18}{20}$ ______________
15. $\frac{3}{14}$ ______________

ALGEBRA Find the missing numerator or denominator.

16. $\frac{2}{8} = \frac{\blacksquare}{24}$
17. $\frac{6}{16} = \frac{\blacksquare}{8}$
18. $\frac{7}{9} = \frac{28}{\blacksquare}$
19. $\frac{2}{5} = \frac{20}{\blacksquare}$

Problem Solving and TAKS Prep

20. Sheryl's cat had a litter of kittens. 3 are white and 6 are gray. What fraction of Sheryl's cat's kittens are white? Write this amount in simplest form.

____

21. Mario ate 4 hot dogs. 1 of them had mustard on it and the rest were plain. What fraction of Mario's hot dogs were plain? Write an equivalent fraction for this amount.

____

22. Which fraction is equivalent to $\frac{2}{5}$?

A $\frac{3}{6}$ C $\frac{4}{10}$

B $\frac{2}{8}$ D $\frac{5}{15}$

23. What is $\frac{15}{40}$ in simplest form?

F $\frac{1}{4}$ H $\frac{3}{8}$

G $\frac{5}{5}$ J $\frac{1}{3}$

Name ______________________

Lesson 15.3

Problem Solving Workshop Strategy: Compare Strategies

Problem Solving Strategy Practice

Draw a picture or act it out to solve each problem.

1. A carousel has a total of 8 horses. 2 of the horses have red saddles. What fraction of horses on the carousel have red saddles?

____

2. On a carousel 4 of the 8 horses have blue saddles. What are 2 equivalent fractions that show the number of horses on the carousel that have blue saddles?

____

3. A student wants to find an equivalent fraction for $\frac{2}{3}$ using a denominator of 9. Solve and explain the strategy you use.

____

____

____

4. A student wants to find two equivalent fractions for $\frac{4}{12}$ using denominators that are less than 12. Solve and explain the strategy you used.

____

____

____

Mixed Strategy Practice

USE DATA For 5–6, use the table below.

Midland Park Antique Carousel	
Animals	**Number on Carousel**
Horse	6
Zebra	4
Elephant	2
Giraffe	3

5. There were twice as many children on the carousel as there were horses. There were 10 more children on the carousel than adults. How many adults were on the carousel? Guess and check.

____

6. What are two equivalent fractions that show what part of the animals on the carousel are giraffes? Draw a picture to show your answer.

____

____

Name____ **Lesson 15.4**

Compare Fractions

Model each fraction to compare. Write <, >, or = for each ●.

1. $\frac{6}{9}$ ● $\frac{8}{9}$ **2.** $\frac{4}{5}$ ● $\frac{2}{3}$ **3.** $\frac{1}{5}$ ● $\frac{1}{8}$

4. $\frac{2}{6}$ ● $\frac{1}{3}$ **5.** $\frac{2}{4}$ ● $\frac{3}{5}$ **6.** $\frac{3}{8}$ ● $\frac{5}{8}$

7. $\frac{3}{5}$ ● $\frac{3}{4}$ **8.** $\frac{1}{3}$ ● $\frac{5}{8}$ **9.** $\frac{3}{8}$ ● $\frac{3}{4}$

10. $\frac{1}{2}$ ● $\frac{1}{3}$ **11.** $\frac{5}{6}$ ● $\frac{5}{8}$ **12.** $\frac{3}{8}$ ● $\frac{4}{8}$

Use number lines to compare.

13. $\frac{3}{5}$ ● $\frac{3}{4}$ **14.** $\frac{5}{9}$ ● $\frac{4}{8}$ **15.** $\frac{4}{10}$ ● $\frac{2}{5}$

16. $\frac{3}{10}$ ● $\frac{3}{8}$ **17.** $\frac{4}{12}$ ● $\frac{1}{5}$ **18.** $\frac{4}{16}$ ● $\frac{6}{12}$

19. $\frac{1}{5}$ ● $\frac{3}{10}$ **20.** $\frac{2}{3}$ ● $\frac{6}{9}$ **21.** $\frac{3}{4}$ ● $\frac{6}{8}$

22. $\frac{2}{6}$ ● $\frac{2}{9}$ **23.** $\frac{5}{8}$ ● $\frac{1}{3}$ **24.** $\frac{2}{4}$ ● $\frac{4}{10}$

25. $\frac{3}{7}$ ● $\frac{4}{7}$ **26.** $\frac{2}{6}$ ● $\frac{2}{8}$ **27.** $\frac{5}{9}$ ● $\frac{9}{12}$

Name______________________________ Lesson 15.5

Order Fractions

Order the fractions from least to greatest. You may use a model.

1. $\frac{1}{3}, \frac{1}{8}, \frac{1}{6}$ ______

2. $\frac{4}{5}, \frac{3}{5}, \frac{5}{8}$ ______

3. $\frac{4}{10}, \frac{4}{12}, \frac{4}{8}$ ______

4. $\frac{3}{7}, \frac{5}{10}, \frac{5}{8}$ ______

5. $\frac{1}{9}, \frac{4}{5}, \frac{2}{3}$ ______

6. $\frac{5}{6}, \frac{6}{10}, \frac{1}{12}$ ______

7. $\frac{5}{12}, \frac{2}{4}, \frac{4}{6}$ ______

8. $\frac{3}{9}, \frac{2}{10}, \frac{5}{6}$ ______

Order the fractions from greatest to least. You may use a model.

9. $\frac{1}{5}, \frac{1}{4}, \frac{1}{8}$ ______

10. $\frac{4}{9}, \frac{4}{5}, \frac{2}{3}$ ______

11. $\frac{3}{4}, \frac{3}{8}, \frac{3}{5}$ ______

12. $\frac{2}{10}, \frac{2}{5}, \frac{3}{12}$ ______

13. $\frac{5}{12}, \frac{3}{9}, \frac{3}{6}$ ______

14. $\frac{7}{12}, \frac{3}{4}, \frac{2}{4}$ ______

15. $\frac{5}{8}, \frac{4}{6}, \frac{1}{10}$ ______

16. $\frac{3}{5}, \frac{6}{12}, \frac{2}{10}$ ______

Problem Solving and TAKS Prep

17. Matt made a fruit salad that included $\frac{3}{4}$ cup of strawberries, $\frac{5}{8}$ cup of grapes, and $\frac{2}{4}$ cup of blueberries. Order the amounts from least to greatest.

______

18. Carolyn walks $\frac{4}{6}$ mile home from school. John walks $\frac{3}{8}$ mile home from school and Corey walks $\frac{6}{12}$ mile home from school. Order their distances from greatest to least.

______

19. Pat spent $\frac{3}{9}$ of her day shopping, $\frac{2}{10}$ of her day exercising, and $\frac{2}{5}$ of her day studying. Which activity took the longest?

______

20. In a jar of marbles, there are $\frac{3}{10}$ red marbles, $\frac{1}{5}$ blue marbles, and $\frac{2}{15}$ white marbles. Of which color are there the least marbles?

______

Name______________________________ **Lesson 15.6**

Read and Write Mixed Numbers

Write a mixed number for each picture.

1. 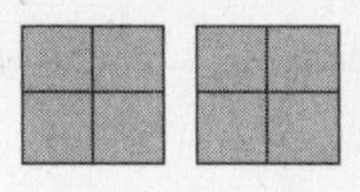

2. 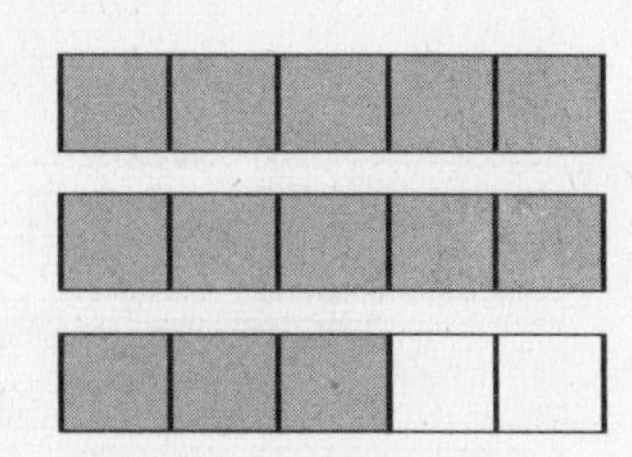

3. 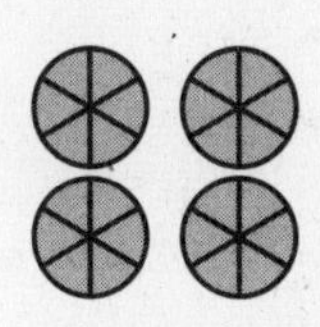

______ ______ ______

For 4–8, use the number line to write the letter each mixed number or fraction represents.

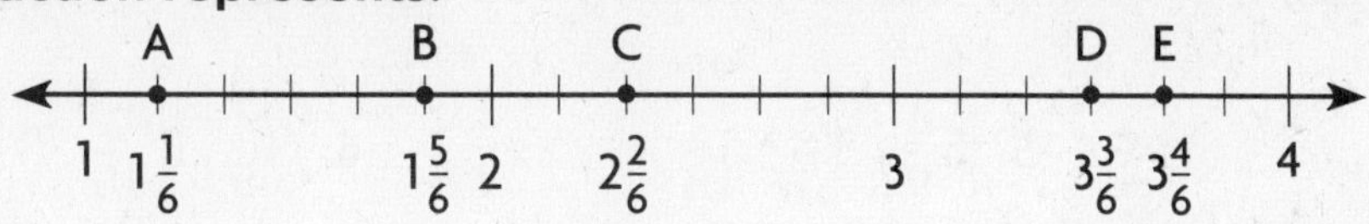

4. $\frac{14}{6}$ ______ 5. $3\frac{4}{6}$ ______ 6. $\frac{11}{6}$ ______ 7. $3\frac{3}{6}$ ______ 8. $\frac{7}{6}$ ______

Rename each fraction as a mixed number and each mixed number as a fraction. You may wish to draw a picture.

9. $5\frac{3}{4}$ 10. $3\frac{2}{10}$ 11. $\frac{38}{6}$ 12. $\frac{23}{3}$ 13. $2\frac{3}{8}$

______ ______ ______ ______ ______

Problem Solving and Test Prep

14. Ned cuts a board that is $5\frac{1}{4}$ inches long. Draw a number line and locate $5\frac{1}{4}$ inches.

15. Julia goes for a bike ride for $1\frac{2}{3}$ hours. Draw a number line to represent the length of time.

16. Denzel makes a cake with $2\frac{2}{3}$ cups of flour. Which shows the mixed number as a fraction?

A $\frac{4}{3}$

B $\frac{8}{3}$

C $\frac{6}{3}$

D $\frac{10}{3}$

17. Ashley serves $3\frac{5}{8}$ trays of muffins. How many muffins does Ashley serve if each muffin is $\frac{1}{8}$ of a tray?

F 29

G 15

H 24

J 19

Name____

Compare and Order Mixed Numbers

Compare the mixed numbers. Use <, >, or =.

1. 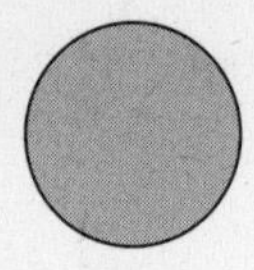

$1\frac{3}{5}$ ● $1\frac{3}{4}$

2. 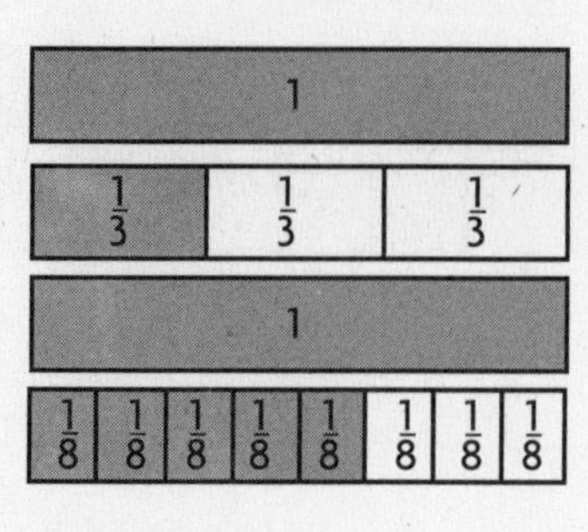

$1\frac{1}{3}$ ● $1\frac{5}{8}$

3. 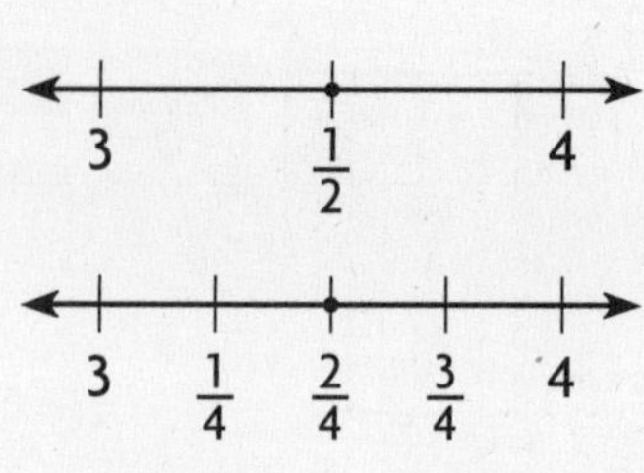

$3\frac{1}{2}$ ● $3\frac{2}{4}$

Order the mixed numbers from least to greatest. You may use a model.

4. $2\frac{1}{4}, 4\frac{3}{8}, 2\frac{3}{4}$

____

5. $5\frac{4}{9}, 5\frac{2}{3}, 5\frac{1}{8}$

____

6. $3\frac{4}{5}, 3\frac{2}{10}, 3\frac{5}{12}$

____

7. $6\frac{3}{6}, 6\frac{3}{4}, 6\frac{1}{3}$

____

8. $1\frac{3}{8}, 1\frac{3}{5}, 1\frac{3}{9}$

____

9. $7\frac{1}{4}, 7\frac{1}{7}, 7\frac{3}{5}$

____

Problem Solving and TAKS Prep

USE DATA For 10–11, use the table.

10. Which ingredient shows the largest amount?

____

11. Which ingredient requires $\frac{5}{3}$ cups?

____

Recipe for Trail Mix	
Ingredient	**Amount**
Corn chips	2 cups
Peanuts	$1\frac{1}{3}$ cups
Raisins	$1\frac{2}{3}$ cups

12. Jamal plays soccer for $\frac{12}{5}$ hours. Write the amount of time Jamal plays soccer as a mixed number.

____

13. Eddie is at an amusement park and wants to find the ride with the shortest wait. The waits for four rides are shown. Which wait is the shortest?

A $1\frac{4}{5}$

B $1\frac{1}{5}$

C $1\frac{1}{2}$

D $1\frac{2}{3}$

Name____ Lesson 16.1

Relate Fractions and Decimals

Write the fraction and decimal shown by each model.

1. 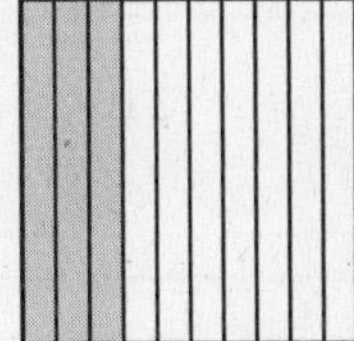
2. 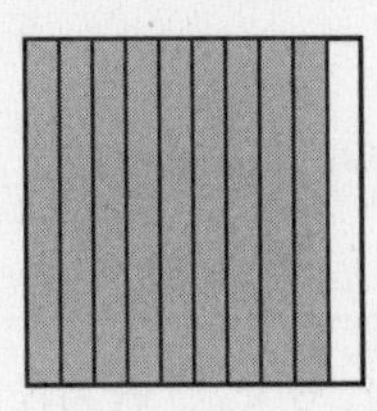
3. 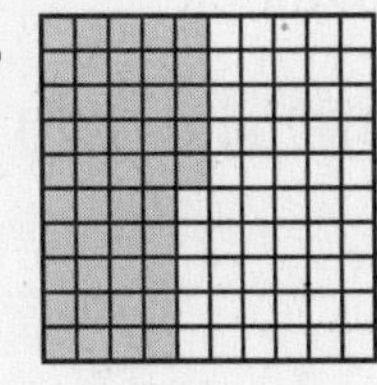
4. 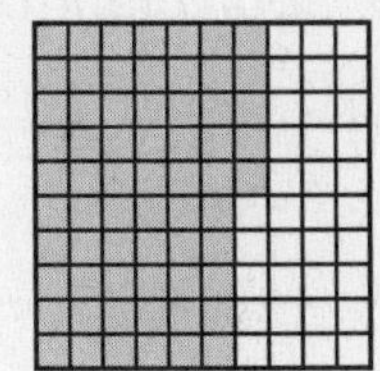

________ ________ ________ ________

Write each fraction as a decimal. You may use a model or drawing.

5. $\frac{6}{10}$
6. $\frac{2}{100}$
7. $\frac{1}{10}$
8. $\frac{63}{100}$

________ ________ ________ ________

ALGEBRA Find the missing number. You may use a model or drawing.

9. 2 tenths + 5 hundredths = ______
10. ______ tenths + 4 hundredths = 0.04
11. 3 tenths + ______ hundredths = 0.38
12. ______ tenths + 0 hundredths = 0.10
13. 9 tenths + 7 hundredths = ______
14. 6 tenths + ______ hundredths = 0.66

Problem Solving and TAKS Prep

15. Write five cents in standard form.

____

16. Write one and thirty-four hundredths in standard form.

____

17. Which fraction does the model represent?

A $\frac{9}{100}$

B $\frac{6}{100}$

C $\frac{9}{10}$

D $\frac{6}{10}$

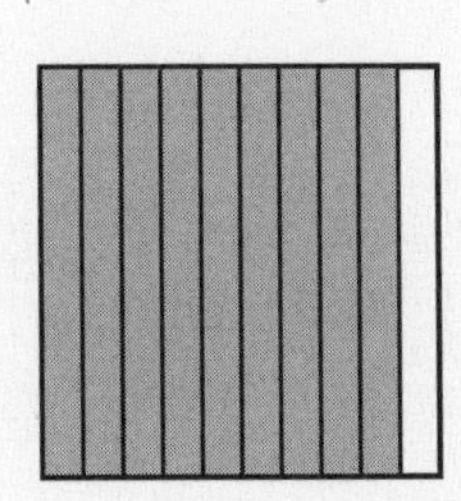

18. Mori ran 0.45 mile on Tuesday. Sharetha ran $\frac{6}{10}$ mile the same day. Sharetha says she ran farther than Mori. Explain how to use a number line to show if Sharetha is correct.

____

____

____

____

Name____ **Lesson 16.2**

Equivalent Decimals

Use a tenths model and a hundredths model. Are the two decimals equivalent? Write *equivalent* or *not equivalent*.

1. 0.2 and 0.02 ____________ **2.** 0.2 and 0.20 ____________ **3.** 0.5 and 0.51 ____________ **4.** 0.6 and 0.06 ____________

5. 0.7 and 0.70 ____________ **6.** 0.11 and 0.1 ____________ **7.** 0.3 and 0.30 ____________ **8.** 0.44 and 0.42 ____________

Write an equivalent decimal for each. You may use decimal models.

9. 0.90 ____________ **10.** 0.5 ____________ **11.** 0.70 ____________ **12.** 0.80 ____________

13. $\frac{4}{10}$ ____________ **14.** $1\frac{1}{2}$ ____________ **15.** $\frac{50}{100}$ ____________ **16.** $3\frac{45}{100}$ ____________

ALGEBRA Write an equivalent decimal. Use the models to help.

17.

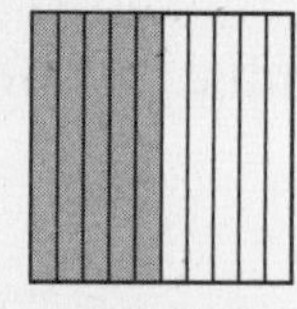 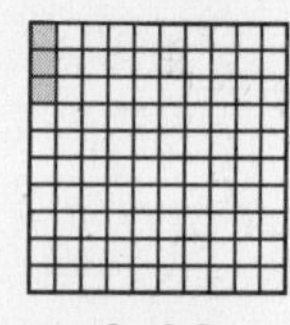 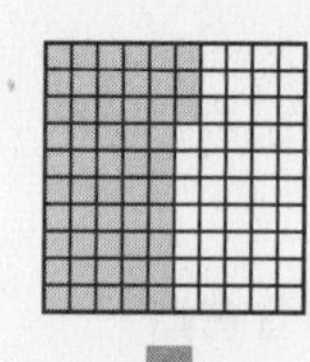

0.5 + 0.03 = ■

18.

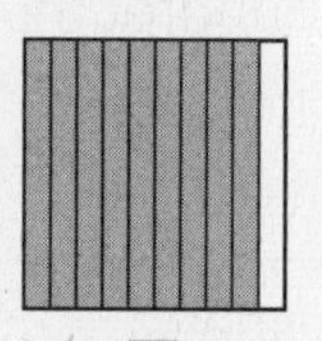 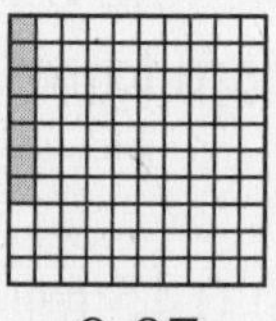 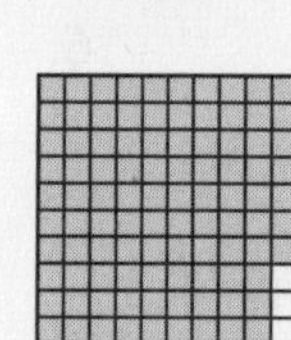

■ + 0.07 = 0.97

19.

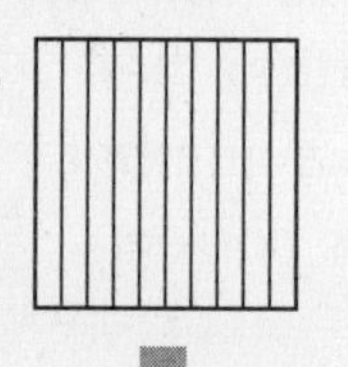 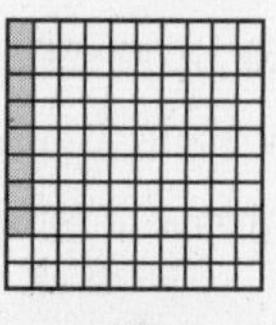 

■ + ■ = 0.08

20.

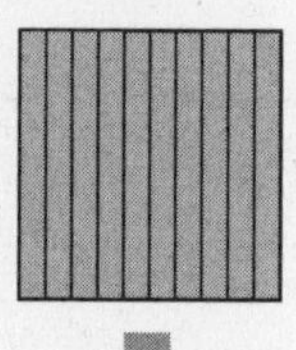 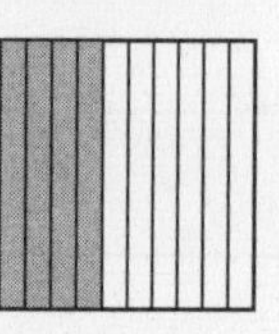 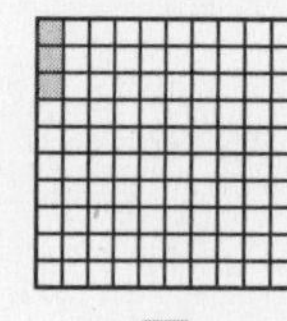 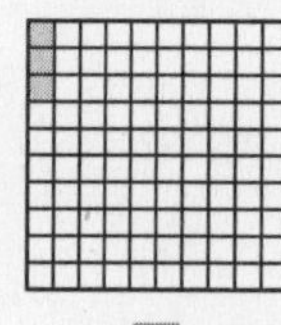 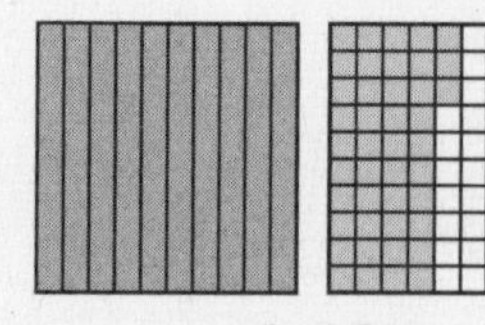

■ + 0.4 + ■ = 1.43

Name____

Relate Mixed Numbers and Decimals

Write a mixed number and an equivalent decimal for each model.

1. 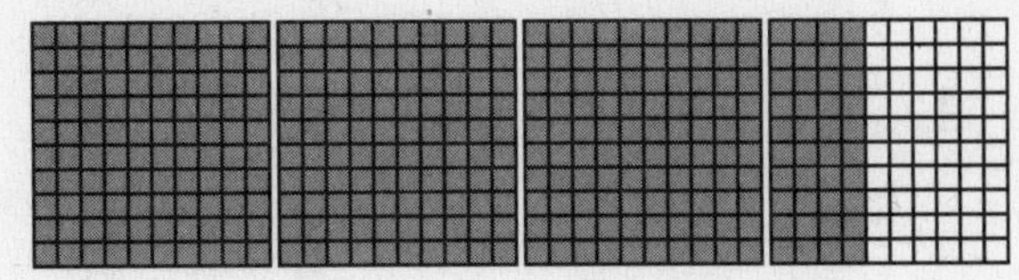

2. 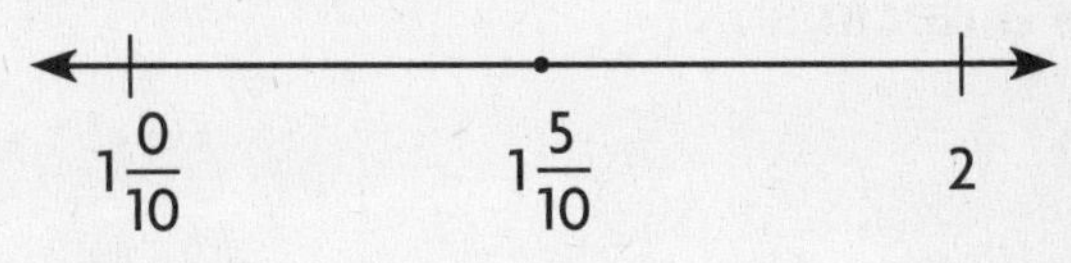

________________________ ________________________

Write an equivalent mixed number and a decimal for each. Then write the word form. You may use a model.

3. 6.6 4. $3\frac{90}{100}$ 5. 4.75 6. $5\frac{1}{4}$ 7. 2.09

ALGEBRA Write the missing number in each ■.

8. $2.4 = 2 + ■$

9. $3.80 = 3 + 0.8 + ■$

10. $5.06 = 5 + ■ + 0.00$

Problem Solving and TAKS Prep

11. Harriet is thinking of a decimal that is equivalent to eight and one-fifth. What is that decimal?

____

12. A CD case measures four and four-fifths inches by five and three-fifths inches. What is the decimal form of the measurements?

____

13. Which mixed number is equivalent to 3.25?

A $3\frac{1}{4}$

B $3\frac{2}{5}$

C $3\frac{2}{3}$

D $2\frac{9}{100}$

14. In simplest form, what is an equivalent fraction for the decimal 2.36?

F $2\frac{4}{50}$

G $2\frac{3}{10}$

H $2\frac{9}{25}$

J $2\frac{6}{100}$

Fractions, Decimals, and Money

Write the total money amount. Then write the amount as a fraction of a dollar and as a decimal.

1. ____________

2.  ____________

3.  ____________

4.  ____________

5. 3 quarters, 1 dime ____________

6. 1 quarter, 3 nickels ____________

7. 5 nickels, 3 dimes ____________

8. 1 dime, 1 nickel, 2 pennies ____________

Write the amount as a fraction of a dollar, as a decimal, and as a money amount.

9. 6 dimes ____________

10. 2 nickels, 7 pennies ____________

11. 4 dimes, 9 pennies ____________

12. 8 dimes, 12 pennies ____________

ALGEBRA Find the missing number to tell the value of each digit.

13. $2.72 = ______ dollars + ______ dimes + ______ pennies

2.72 = ______ ones + ______ tenths + ______ hundredths

14. $8.06 = ______ dollars + ______ pennies

8.06 = ______ ones + ______ hundredths

Problem Solving and TAKS Prep

USE DATA For 15–16, use the table.

15. What fruit could $\frac{1}{2}$ of a dollar buy?

____

16. What fruit could $\frac{3}{4}$ of a dollar buy?

____

School Cafeteria	
Fruit	**Price**
Apple	$0.35
Banana	$0.45
Pear	$0.55

17. What decimal do 4 dimes represent?

A 0.35 **C** 0.41

B 0.40 **D** 0.45

18. What decimal do 2 quarters represent?

____

Name____ **Lesson 16.5**

Compare Decimals

Compare. Write $<$, $>$, or $=$ for each ●.

1.

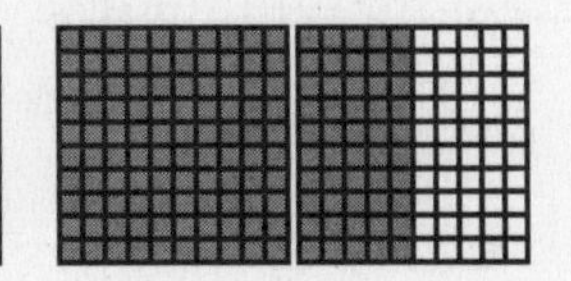

1.51 ● 1.5

2.

0.30 ● 0.3

3.

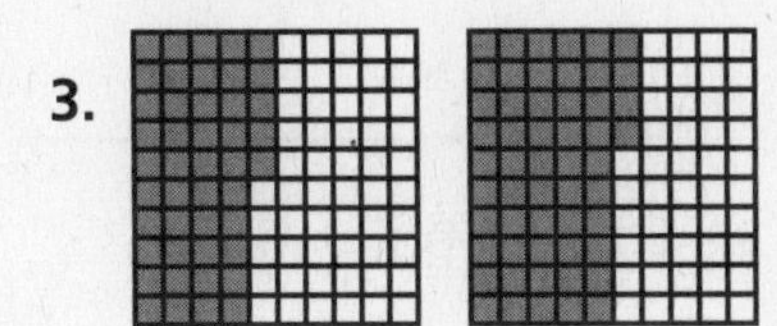

0.45 ● 0.54

4.

1.20 ● 1.02

5.

2.09 ● 2.90

6.

2.34 ● 1.43

Use the number line to determine whether the number sentences are *true* or *false*.

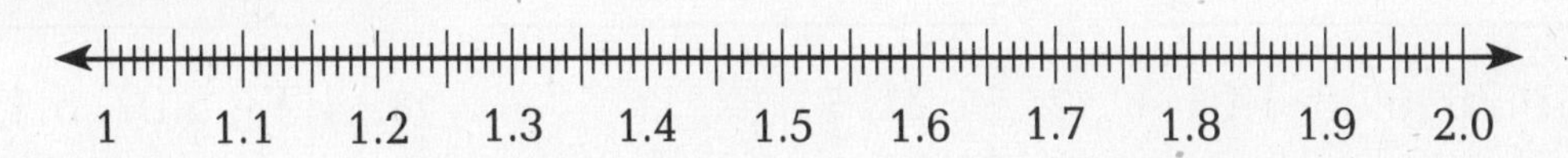

7. $1.25 < 1.52$ ____________

8. $1.70 > 1.7$ ____________

9. $1.21 < 1.2$ ____________

10. $1.22 < 1.11$ ____________

11. $1.29 < 1.92$ ____________

12. $1.4 = 1.40$ ____________

13. $1.09 > 1.08$ ____________

14. $1.66 = 1.67$ ____________

15. $1.37 < 1.35$ ____________

16. $1.55 > 1.45$ ____________

17. $1.0 = 1.00$ ____________

18. $1.9 < 1.99$ ____________

Name___________________________________

Lesson 16.6

Order Decimals

Use the number line to order the decimals from least to greatest.

1 1.1 1.2 1.3 1.4 1.5 1.6 1.7 1.8 1.9 2.0

1. 1.45, 1.44, 1.43

____

2. 1.05, 1.04, 1.4

____

3. 1.78, 1.79, 1.09

____

4. 1.33, 1.32, 1.3

____

5. 1.2, 1.19, 1.27

____

6. 1.05, 1.03, 1.01

____

7. 1.02, 1.03, 1.1

____

8. 1.84, 1.89, 1.82

____

9. 1.66, 1.65, 1.62

____

Order the decimals from greatest to least.

10. 1.66, 1.06, 1.6, 1.65

____

11. \$5.33, \$5.93, \$5.39, \$3.55

____

12. 4.84, 4.48, 4.88, 4.44

____

13. \$1.45, \$1.43, \$1.54, \$1.34

____

14. 7.32, 7.38, 7.83, 7.23

____

15. \$0.98, \$1.99, \$0.89, \$1.89

____

16. 0.67, 0.76, 0.98, 1.01

____

17. \$1.21, \$1.12, \$1.11, \$1.10

____

18. 4.77, 5.07, 5.1, 4.6

____

19. 1.21, 1.45, 1.12, 1.44

____

20. 2.21, 2.67, 2.66, 2.3

____

21. \$9.00, \$9.10, \$9.11, \$9.99

____

22. \$5.97, \$5.96, \$6.59, \$5.75

____

23. \$3.39, \$3.03, \$3.83, \$3.30

____

24. 8.17, 8.05, 8.08, 8.1

____

Name________________________________ Lesson 16.7

Problem Solving Workshop Skill: Draw Conclusions

Problem Solving Skill Practice

Use the information on the chart to draw a conclusion

1. Janae looks at the ads to the right and wants the best value for her money. If she wants one game, which one should she buy at which store?

____

2. What if Great Games sold playing cards for $3.50. Which store would have the better value?

____

Mixed Applications

USE DATA For 3–4, use the map.

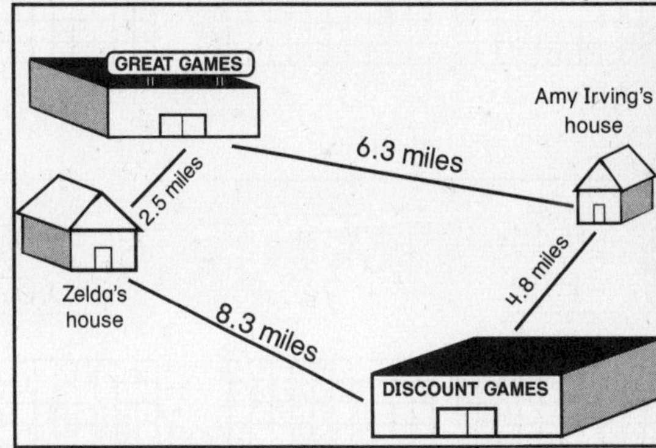

3. Sal lives 4.08 miles from Discount Games. Who lives closer, Amy or Sal?

____

4. If Sal lives 6.33 miles from Great Games, who lives closest to each store? List Amy, Sal, and Zelda in the order they live from each store from least to greatest.

____

____

5. Patty paid for 8 Tee-shirts with five $20-dollar bills. If the tees each cost $12.35, about how much change did Patty get back? Do you need an estimate or an exact answer?

____

____

Name ______________________

Model Addition

Use models to find the sum.

1. $\begin{array}{r} 0.56 \\ +0.45 \\ \hline \end{array}$

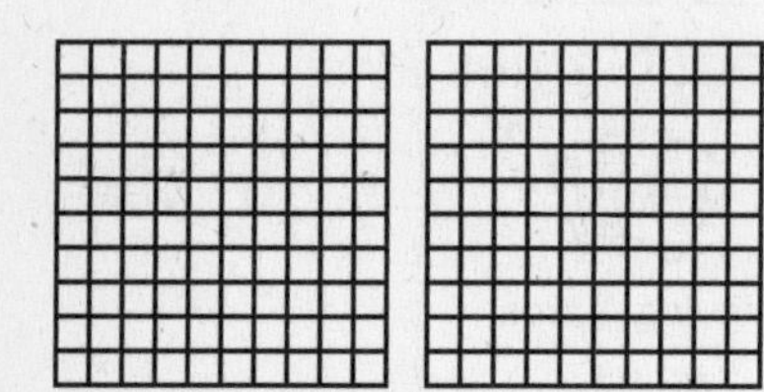

______

2. $\begin{array}{r} 0.4 \\ +0.7 \\ \hline \end{array}$

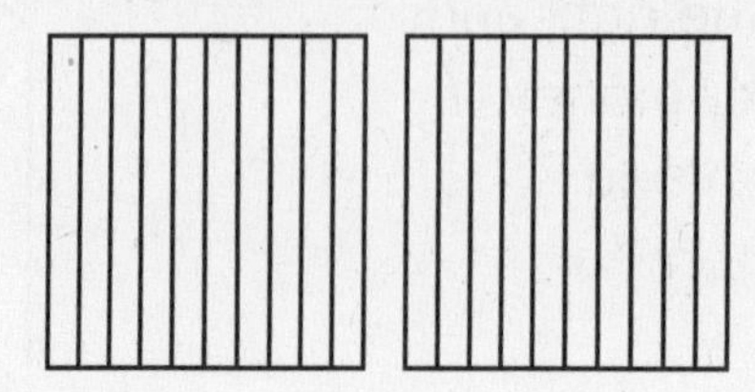

______

3. $\begin{array}{r} 0.25 \\ +0.07 \\ \hline \end{array}$

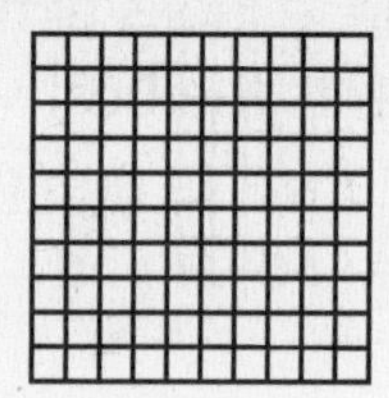

______

4. $\begin{array}{r} 1.05 \\ +0.78 \\ \hline \end{array}$

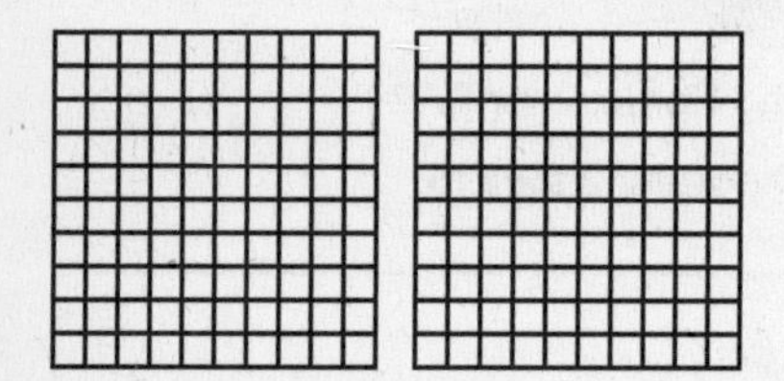

______

5. $\begin{array}{r} 0.38 \\ +1.93 \\ \hline \end{array}$

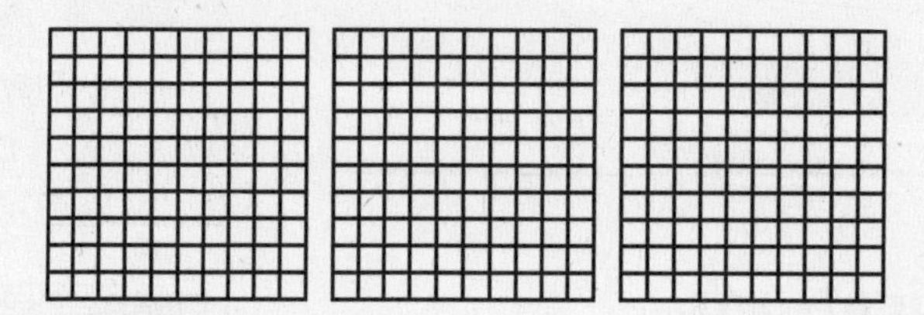

______

6. $\begin{array}{r} 0.44 \\ +1.08 \\ \hline \end{array}$

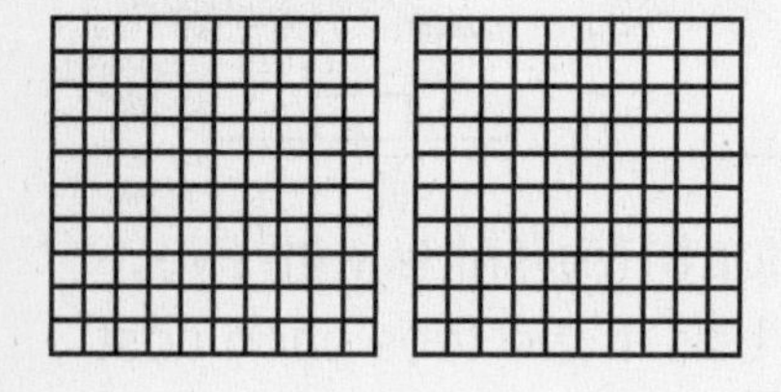

______

7. $\begin{array}{r} 1.06 \\ +0.67 \\ \hline \end{array}$

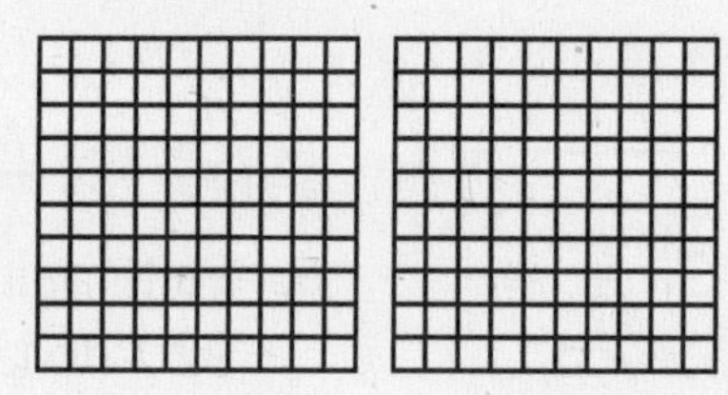

______

8. $\begin{array}{r} 0.16 \\ +1.55 \\ \hline \end{array}$

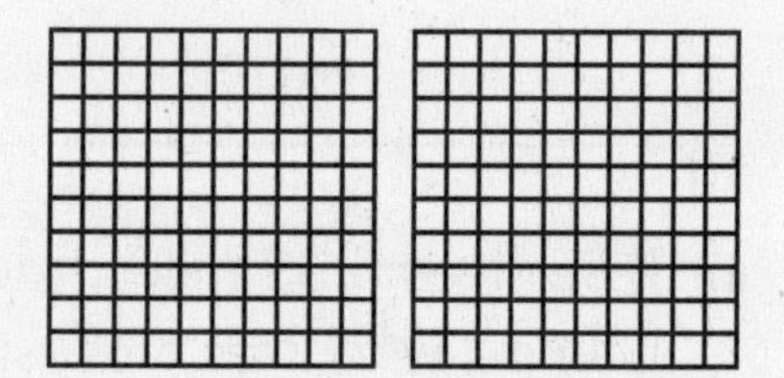

______

ALGEBRA Use the models to find the missing addend.

9. 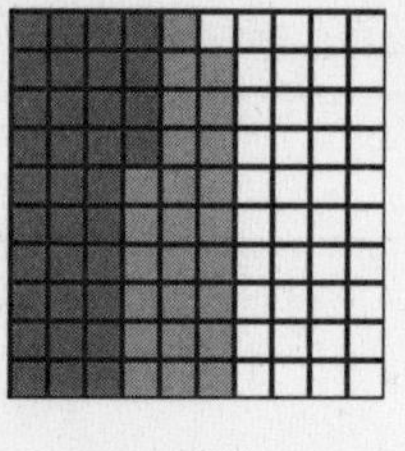

0.7 + ■ = 0.9

10. 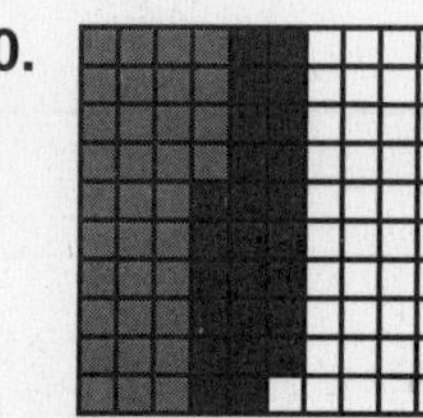

0.34 + ■ = 0.59

Name____ Lesson 17.2

Model Subtraction

Use models to find the difference.

1. $\begin{array}{r} 0.57 \\ -0.18 \\ \hline \end{array}$

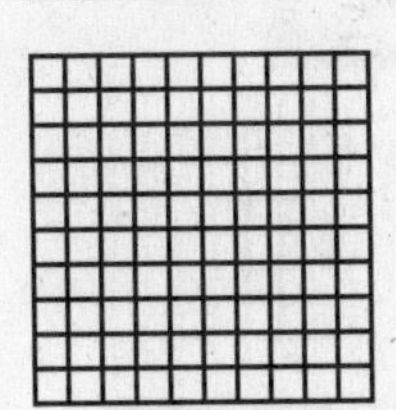

2. $\begin{array}{r} 0.7 \\ -0.3 \\ \hline \end{array}$

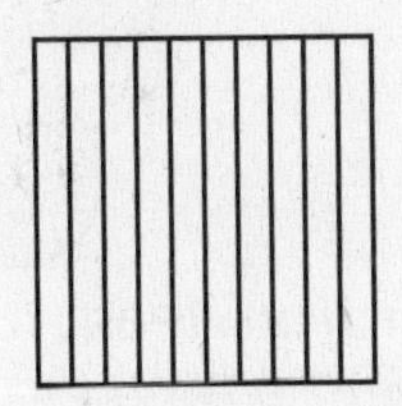

3. $\begin{array}{r} 1.07 \\ -0.42 \\ \hline \end{array}$

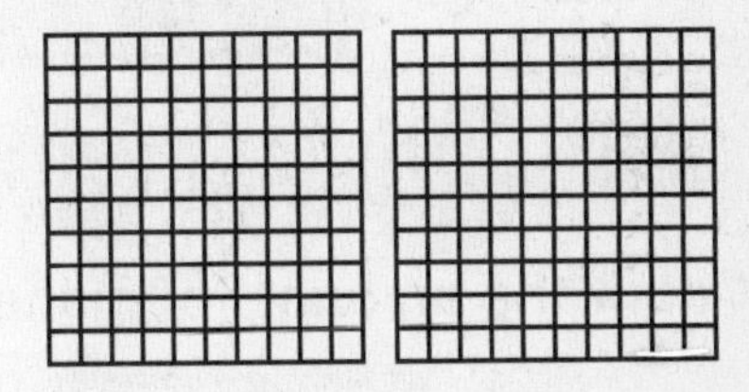

4. $\begin{array}{r} 1.09 \\ -0.90 \\ \hline \end{array}$

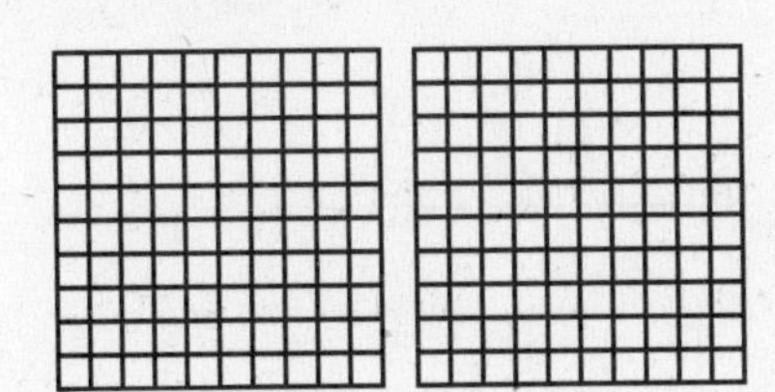

5. $\begin{array}{r} 1.00 \\ -0.63 \\ \hline \end{array}$

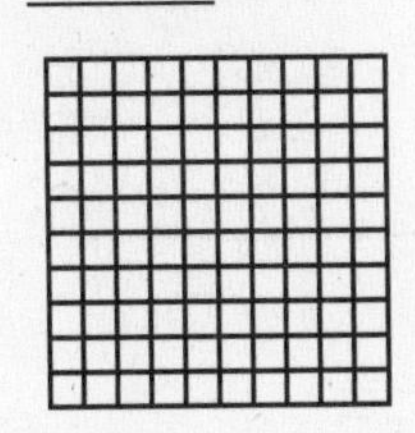

6. $\begin{array}{r} 1.98 \\ -1.29 \\ \hline \end{array}$

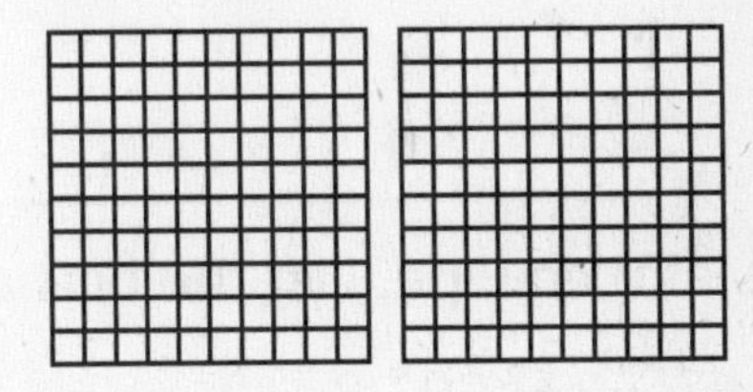

7. 2.73 – 1.79

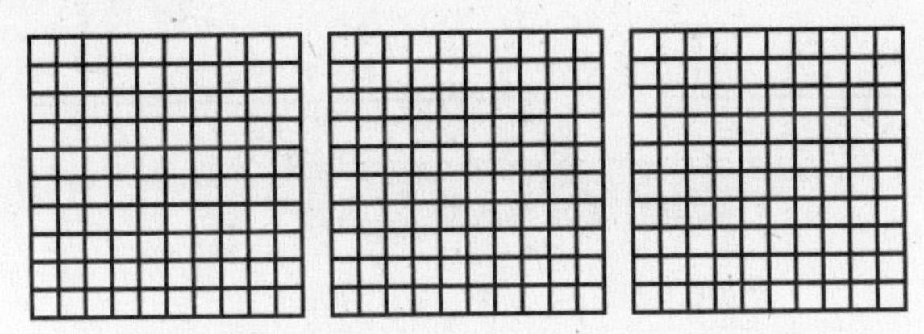

____

8. 2.92 – 2.07

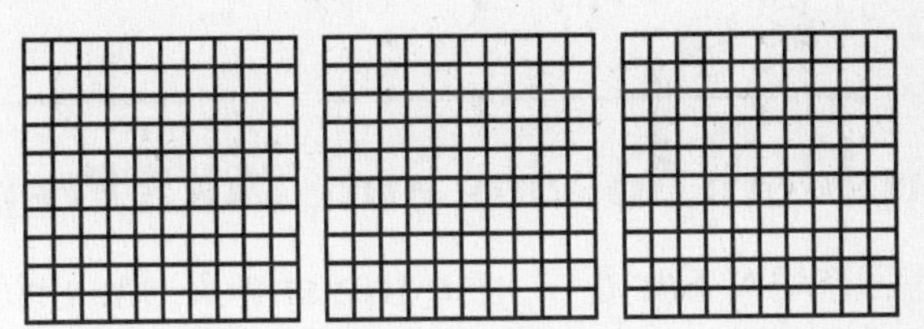

____

ALGEBRA Use the models to find the missing number.

9. 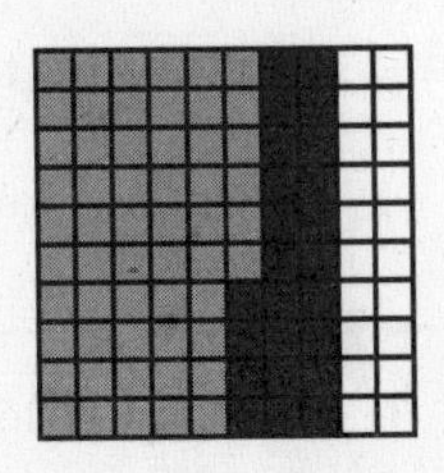

0.80 – ■ = 0.16

10. 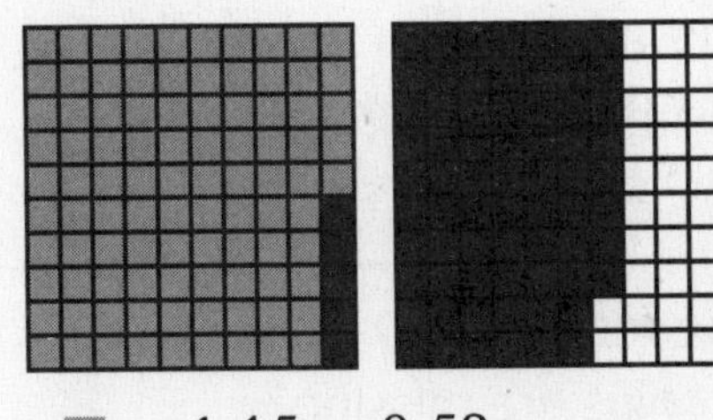

■ – 1.15 = 0.53

Name ______________________________

Use Money

Compare. Write <, >, or =.

1. a.

b.

Count the amounts of money and order from least to greatest.

2.

____

Make change. List the bills and coins.

3. Cost $0.38

Paid with: 

____

4. Cost $7.52

Paid with: 

____

5. Cost $19.13

Paid with: 

____

Problem Solving and TAKS Prep

USE DATA For 6–7, use the snack bar prices.

Beach Snack Bar	
Food Item	**Price**
Salad	$1.75
Sandwich	$3.29
Yogurt	$0.99
Fruit Cup	$1.45
Pizza	$1.25 per slice

6. Bart bought 6 pizza slices and 3 salads. If he paid for it all with a $20 bill, how much change did Bart get?

____

7. Faith bought one of each item on the menu. How much change did Faith get if she paid with a $10 bill?

____

8. Charlie received $0.62 in change. What coins are missing?

____

9. Jay received $0.87 in change. What coins are missing?

____

Name________________________________

Lesson 17.4

Add and Subtract Decimals and Money

Use the models to find the sum or difference. Record the answer.

1. $0.7 - 0.3$

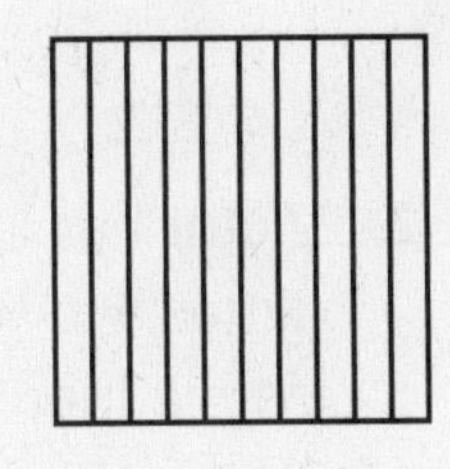

____

2. $0.44 + 1.08$

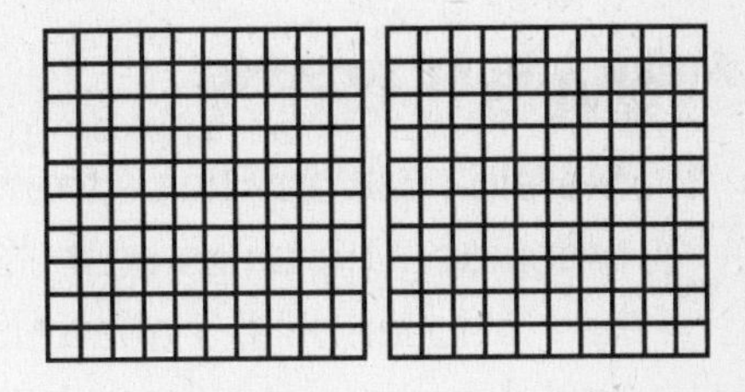

____

3. $2.92 - 2.07$

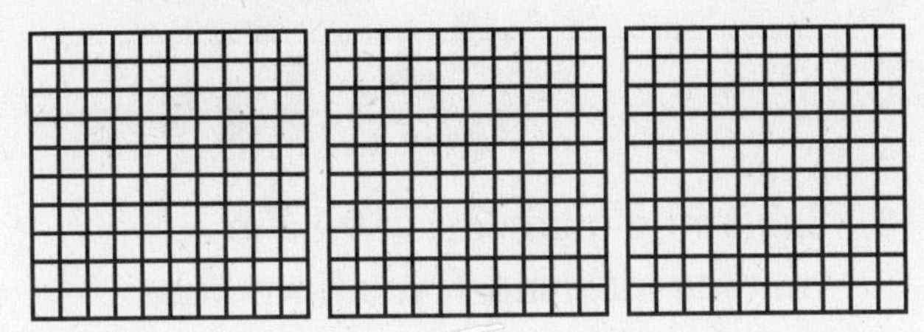

____

4. $0.22 + 0.95$

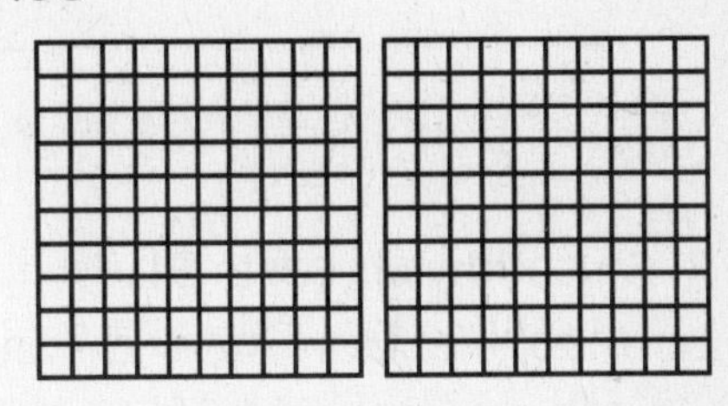

____

Compare. Write <, >, or = for each ●.

5. $5.15 + $0.10 ● $4.84 + $0.35

6. 3.78 + 2.51 ● 9.54 − 3.30

ALGEBRA **Find the missing decimals. The sums are given at the end of each row and the bottom of each column.**

7.	13.06	4.12	22.77	
8.	67.77		15.14	83.64
9.	0.98	73.22		80.78
10.		78.07	44.49	

Problem Solving and TAKS Prep

11. Lyle spent $2.47 on peanut butter, $3.56 on jelly, and $2.37 on a loaf of bread. How much did he spend in all?

____

12. Jason bought pants on sale for $25.89. This was $8.09 less than the original price. Use a model or draw a picture to find the original price of the pants.

A $33.98

B $27.46

C $17.80

D $33.40

Name____ Lesson 17.5

Problem Solving Workshop Strategy: Make a Table

Problem Solving Strategy Practice

Make a table to solve.

1. Maraina will buy chips from a vending machine. The chips cost $2.45. Daina has 2 dollar bills, 3 quarters, 3 dimes, and 4 nickels. What are two different ways Daina can pay for the chips?

____

2. Victor has $1, 4 quarters, and 2 dimes. He will borrow some money from a friend to buy a bag of chips for $2.45. What coin or coins must Victor borrow in order to pay for the chips?

____

3. A sandwich costs $1.00 in a vending machine. How many different ways can you make $1.00 if you have a quarter, some dimes, and some nickels?

____

4. Sugar-free gum costs $0.85 in a vending machine. If you have one quarter, how many dimes would you need to buy a pack of sugar-free gum?

____

Mixed Applications

USE DATA For 5–6, use the table.

5. Tanya spent $9.80 at the pool. What did Tanya pay for?

____

Community Center Pool	
Item	**Price**
Entrance Fee	$1.50
Bathing Cap	$2.75
Towel	$5.55

6. Libby paid for herself and two sisters to go to the pool. She also bought 3 towels and a bathing cap. How much did Libby spend?

____

7. Henry had the exact change to pay for a $0.50 pencil. He paid with 6 coins. What could those coins be?

____

8. In Exercise 1, how much money will Dana have left over after she buys the chips?

____

Name______________________________

Lesson 18.1

Measure Time

Write the time as shown on a digital clock and an analog clock.

1. two fifty-two ____________

2. 12 minutes after eight ____________

3. half past six ____________

Write two ways to read the time. Then estimate to the nearest 5 minutes.

4. 

____

____

5. 4:16 A.M.

____

____

6. 6:47 P.M.

____

____

7. 

____

____

8. 

____

____

9. 

____

____

Tell whether to use seconds, minutes, hours, or days to measure time.

10. to travel around the world ____________

11. to look up the meaning of a word ____________

12. to make a model plane from scratch ____________

Problem Solving and TAKS Prep

Use the watch.

13. To the nearest 5 minutes, what time is it?

14. Write the time in two other ways.

____

15. Jose has lunch at quarter after eleven. What time is it?

A 10:45 A.M. **C** 11:30 A.M.

B 11:15 A.M. **D** 11:45 A.M.

16. Greg says that it is eleven minutes until three in the afternoon. What time is it?

F 3:11 P.M. **H** 2:49 P.M.

G 2:49 A.M. **J** 3:11 A.M.

Name______________________________________ **Lesson 18.2**

Elapsed Time

Use a clock to find the elapsed time.

1. start: 8:15 A.M.
end: 8:55 A.M.

____

2. start: 6:50 P.M.
end: 7:20 P.M.

____

3. start: 7:35 A.M.
end: 7:15 P.M.

____

4. start: 9:55 A.M.
end: 1:45 P.M.

____

Find the start time.

5. end: 11:35 P.M.
elapsed time:
6 hr 55 min

____

6. end: 6:25 A.M.
elapsed time:
55 min

____

7. end: 11:41 A.M.
elapsed time:
2 hr 12 min

____

8. end: 8:15 P.M.
elapsed time:
12 hr 25 min

____

9. end: 11:35 A.M.
elapsed time;
3 hr 5 min

____

10. end: 6:12 A.M.
elapsed time:
7 hr 3 min

____

11. end: 9:25 A.M.
elapsed time:
1 hr 50 min

____

12. end: 11:50 A.M.
elapsed time:
5 hr 20 min

____

Problem Solving and TAKS Prep

For 13–14, use the table.

Time Spent at Bus Stops	
Stop	**Elapsed Time (min:sec)**
Avery School	2:05
Central Mall	3:15
Library	1:34
Post Office	1:12

13. At which stop did the bus spend the most time?

____

14. The bus arrived at the library at 3:12 P.M. To the nearest second, when did it leave the library?

____

15. The basketball team starts practice right after the school day ends. They finish practice at 6:00 P.M., which is 2 hours and 30 minutes after the school day ends. What time did the school day end?

A 6:30 P.M.

B 4:00 P.M.

C 3:30 P.M.

D 2:30 P.M.

16. Ms. Smith's social studies class starts at 10:15 A.M. and ends at 11:20 A.M. How long is the class?

F 1 hr

G 1 hr 5 min

H 1 hr 15 min

J 1 hr 20 min

Name________________________________ **Lesson 18.3**

Problem Solving Workshop Skill: Sequence Information

Problem Solving Skill Practice

Exhibit Schedule		
Stop	**Exhibit Times**	**Length**
Light Show	8:00 A.M., 8:30 A.M., 9:00 A.M., 9:30 A.M., 12:00 P.M., 12:30 P.M., 2:00 P.M.	30 minutes
Ecosystems	10:00 A.M., 1:30 P.M.	60 minutes
Claymation	9:00 A.M., 10:00 A.M., 11:30 A.M., 2:00 P.M.	45 minutes
Story Time	8:30 A.M., 9:30 A.M., 11:30 A.M., 3:30 P.M.	30 minutes
Lunch	11:00 A.M., 11:20 A.M., 12:00 P.M., 12:20 P.M., 12:40 P.M., 1:00 P.M.	20 minutes

1. Clay visits the museum from 10:00 A.M. to 12:00 P.M. Which exhibits can Clay see?

____

2. Kara finishes lunch at the museum at 12:20 P.M. She wants to see Claymation next. How long will she have to wait?

____

Mixed Applications

For 3–4, use "Today's Specials".

TODAY'S SPECIALS	
T-Shirts	$10 each
Buy 2 and receive	$2 off
Buy 3 and receive	$5 off
CDs	3 for $21
Bag of Beads	$3 per bag
Bottled Water	$1 each

3. Order the cost *per item* from least to most expensive. Do not include T-shirts in your list.

____

4. Dana spends $38. Marge buys 2 T-shirts, 3 CDs, 3 bags of beads, and 3 bottles of water. How much more does Marge spend than Dana?

____

5. Don kicks a ball 20 feet. Shelly kicks it 2 feet longer than 3 times as far as Don. How far does Shelly kick the ball?

____

6. Jen walks 5 blocks north, 1 block east, and 3 more blocks north. Then she walks 1 block west and 1 block south. How far is Jen from where she started?

____

Name____ Lesson 18.4

Elapsed Time on a Calendar

May						
Sun	Mon	Tue	Wed	Thu	Fri	Sat
				1	2	3
4	5	6	7	8	9	10
11	12	13	14	15	16	17
18	19	20	21	22	23	24
25	26	27	28	29	30	31

June						
Sun	Mon	Tue	Wed	Thu	Fri	Sat
1	2	3	4	5	6	7
8	9	10	11	12	13	14
15	16	17	18	19	20	21
22	23	24	25	26	27	28
29	30					

1. About how many weeks are there between May 13 and June 26?

____

2. Lyle practices for 16 days to perform in a recital on June 8. When did Lyle start practicing?

____

3. Ginger has an appointment May 16. Today is May 5. How many days until Gingers appointment?

____

4. Flag Day is June 14. Todd is going on vacation 3 weeks and 3 days before Flag Day. When is Todd going on vacation?

____

5. Beginning May 28, there is a sale at the department store. The sale lasts for 12 days. What is the last date of the department store's sale?

____

Problem Solving and TAKS Prep

For 6–9, use the calendars above.

6. How many days are there between National Teacher's Day (May 6) and Memorial Day (May 26)?

____

7. National Teacher's Day is always the first Tuesday in May. Explain why it will not always fall on May 6.

____

____

____

8. Memorial Day is May 26. If today is May 15, how many days is it until Memorial Day?

A 15 C 26

B 11 D 16

9. Mrs. Greer returned home from a trip on June 29. If she left on Flag Day (June 14), how many days was Mrs. Greer gone?

F 15 H 2

G 16 J 12

Name______________________________ Lesson 18.5

Temperature: Fahrenheit

Use the thermometer to find the temperature in °F.

1. 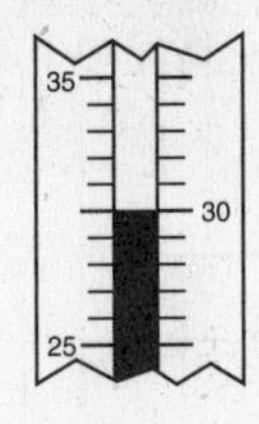

2. 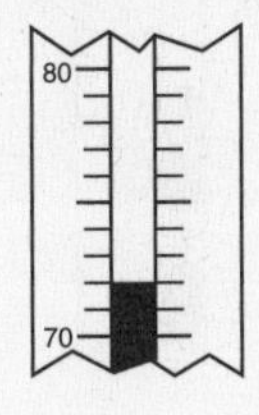

3. 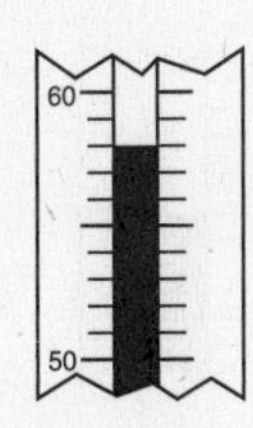

4. 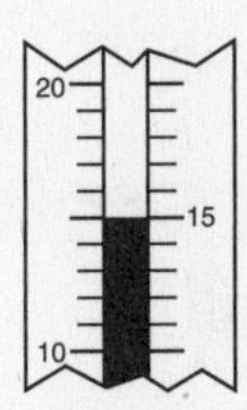

__________ __________ __________ __________

ALGEBRA Use a thermometer to find the change in temperature.

5. 75°F to 58°F
6. 35°F to 47°F
7. ⁻8°F to ⁻25°F
8. 10°F to ⁻9°F

__________ __________ __________ __________

Problem Solving and TAKS Prep

For 9–10, use the table and thermometer.

9. Use the thermometer at the right to label the January and July temperatures in Fairbanks. Find the difference between these temperatures.

____

Average Temperatures		
City	January	July
Baltimore, MD	32°F	77°F
Detroit, MI	25°F	74°F
Fairbanks, AK	⁻10°F	62°F
Madison, WI	17°F	72°F

10. Which city has the least difference between its January and July temperatures?

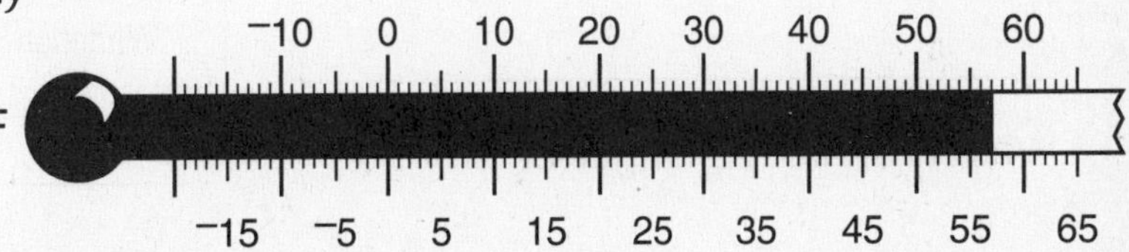

____

11. What is the change in temperature from 45°F to 97°F?

   A 52
   B 55
   C 147
   D 145

12. What is the change in temperature from 3°F to 15°F?

   F 18
   G 45
   H 12
   J 5

Name____ **Lesson 18.6**

Temperature: Celsius

Use the thermometer to find the temperature in °C.

1. 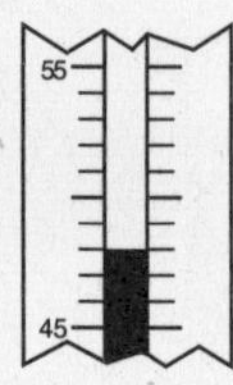

2. 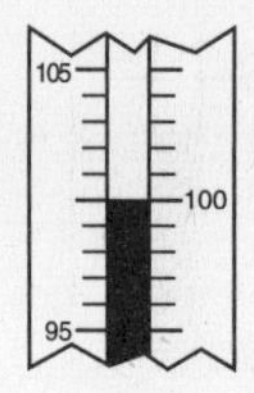

3. 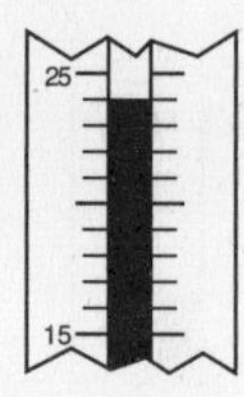

4. 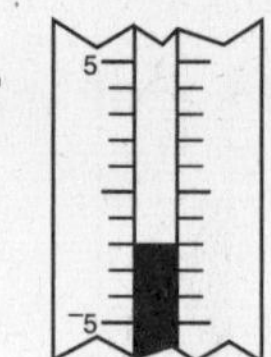

____________ ____________ ____________ ____________

ALGEBRA Use a thermometer to find the change in temperature.

5. 35°C falls ■ to 26°C

6. 18°C rises ■ to 23°C

7. 11°C falls ■ to -6°C

Write a reasonable temperature for each.

8. a bowl of soup

____

9. a summer day

____

10. cold milk for cereal

____

Problem Solving and TAKS Prep

For 11–12, use the table.

11. What change in temperature is necessary to keep Whitefish safe for 4 months rather than 1 month?

____

12. Which fish keeps longer at -29°?

____

Seafood Safe Storage Times			
Type of Fish	**9°C**	**-21°C**	**-29°C**
Whitefish	1 month	4 months	8 months
Herring	1 month	3 months	6 months

13. A temperature of 12°C falls -13°C. What is the new temperature?

____

14. A temperature of -5°C rises 25°. What is the new temperature?

____

Name ____________________ Lesson 19.1

Collect and Organize Data

For 1–2, use the Favorite Snacks frequency table. Tell whether each statement is true or false. Explain.

Students' Favorite Snacks	
Snack	**Votes**
Apple	12
Banana	7
Carrots	8
Celery	4

1. More students chose carrots than bananas.

____

2. More students chose carrots and celery than apples and bananas.

____

For 3–5, use the Sports Participation frequency table.

Sports Participation		
Sport	**Boys**	**Girls**
Golf	12	19
Softball	18	17
Tennis	9	11
Volleyball	13	12

3. How many more boys participate in volleyball than tennis?

____

4. How many more girls participate in golf than in tennis?

____

5. How many more boys and girls together play softball than volleyball?

____

Problem Solving and TAKS Prep

USE DATA For 6–7, use the Sports Participation table above.

6. Which is the most popular sport for girls? For boys?

____

7. Who has the largest overall participation in sports: girls or boys?

____

8. How many people were surveyed in all?

A 186
B 194
C 196
D 200

Favorite Sport	Votes
Golf	37
Softball	63
Tennis	52
Volleyball	44

9. What question would you ask if you were taking a survey about favorite sports?

____

____

Name____

Sort Data

For 1–4, use the Multiples Venn diagram.

1. What labels should you use for sections B and C?

____

2. Why are the numbers 20 and 40 sorted in the B section of the diagram?

____

3. In which section would you sort the number 60? Explain.

____

Multiples

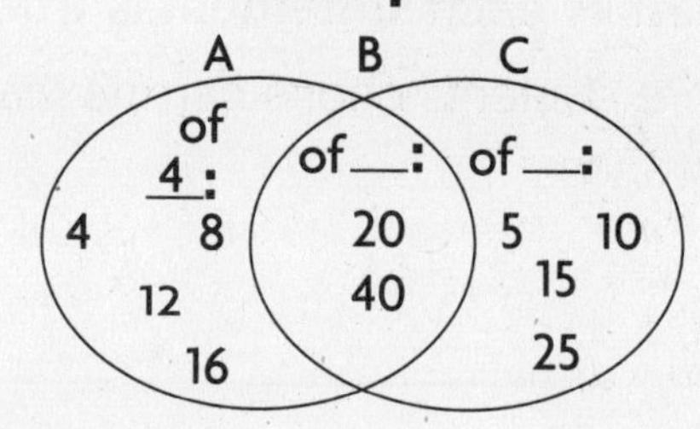

4. **Reasoning** If section A were multiples of 45 and section C were multiples of 71, would section B contain a number less than 100? Explain.

____

For 5–6, use the Breakfast Choices table.

5. Show the results in the Venn diagram at the right.

____

6. What data overlaps? Explain.

____

Breakfast Choices	
Food	**Student Names**
Cereal	Jane, Mani, Liddy, Steve, Ana
Fruit	Ben, Cecee, Beth
Both	Dave, Raiza

Problem Solving and TAKS Prep

USE DATA For 7–8, use the Breakfast Choices table.

7. How many students chose either cereal or fruit?

____

8. What would have to change in order for there to be no overlapping data?

____

____

9. Look at the Multiples Venn diagram at the top of the page. Which number belongs in section C?

   A 22   **C** 204
   B 28   **D** 250

10. Look at the Multiples Venn diagram at the top of the page. Which number belongs in section B?

   F 30   **H** 80
   G 50   **J** 65

Name____________________

Make and Interpret Pictographs

For 1–3, use the Favorite Souvenirs table.

1. How many more tourists chose books than statues?

____

Tourist's Favorite Souvenirs	
Souvenir	Votes
Books	21
Wall hangings	10
Statues	14
Pens and pencils	14

2. How many tourists voted in the survey in all?

____

3. Use the Favorite Souvenirs table to make a pictograph below.

Favorite Souvenirs	
Souvenir	Number of Votes
Books	
Wall hangings	
Statues	
Pens and pencils	

Key: Each ☐ = ☐ votes.

For 4–5, use the Busloads of School Visitors pictograph.

4. Which day had the most busloads of school visitors? the least?

____

5. How many more busloads visited on Thursday and Friday than on Tuesday and Wednesday?

____

Busloads of School Visitors	
Day	Number
Mon	🚌
Tues	🚌🚌🚌
Wed	🚌🚌🚌🚌🚌
Thurs	🚌🚌🚌🚌🚌🚌🚌
Fri	🚌🚌🚌

Key: each 🚌 = 1 busload.

Problem Solving and TAKS Prep

6. What question could you ask for the Favorite Souvenirs Survey?

____

7. In the Busloads pictograph, which days had an equal number of visitors?

____

8. Look at the Favorite Souvenirs pictograph. How many more Tourists chose pens and pencils than wall hangings?

   A 24 **C** 14
   B 4 **D** 10

9. Look at the Busloads pictograph. How many busloads came on the day with the least visitors?

   F 5 **H** 3
   G 7 **J** 1

Name____ Lesson 19.4

Choose a Reasonable Scale

For 1–2 choose 5, 10, or 100 as the most reasonable interval for each set of data. Explain your choice.

1. 35, 55, 77, 85, 20, 17

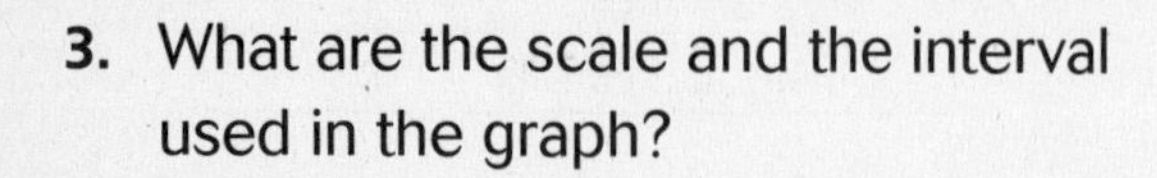

2. 125, 200, 150, 75, 277, 290

____

For 3–6, use the Summer Sport graph.

3. What are the scale and the interval used in the graph?

____

4. How would the length of the bars change if the interval were 10?

____

5. How many votes were cast?

____

6. How many more votes did swimming get than croquet and volleyball combined?

____

Problem Solving and TAKS Prep

USE DATA For 7–10, use the Winter Sport graph.

7. What is the least favorite winter sport?

____

8. How many fewer people voted for sledding than skiing and ice skating combined?

____

9. What is the interval on the Winter Sport graph?

   A 5 **C** 15
   B 10 **D** 20

10. What is the scale of the Winter Sport graph?

   F 0–80 **H** 0–100
   G 0–50 **J** 0–20

Name________________________________ **Lesson 19.5**

Interpret Bar Graphs

For 1–6, use the Distance of Planets bar graph.

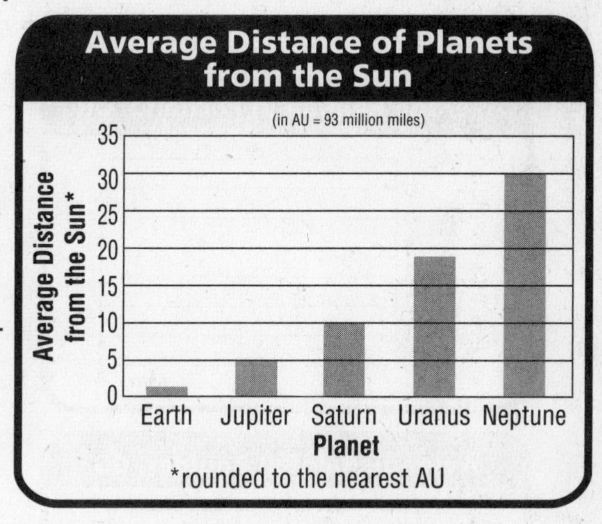

1. An Astronomical Unit (AU) is the average distance between the Earth and the sun. Scientists use Astronomical Units to help represent other large distances. According to the data shown in the graph, what is the range of AU shown?

   ____

2. Which planet in the graph is farthest from the sun?

   ____

3. Which planet is 6 times farther away from the Sun than Jupiter?

   ____

4. Which planet's distance from the Sun is the median of the data?

   ____

5. List the names of the planets in the graph in order from the greatest average distance from the Sun to the least average distance from the Sun.

   ____

   ____

   ____

6. **Reasoning** Of the planets shown in the graph, which planet do you think is the coldest? Which planet do you think is the warmest? Why?

   ____

   ____

   ____

Problem Solving and TAKS Prep

USE DATA For 7–10, use the Distance of Planets bar graph above.

7. How many AU longer is Uranus' average distance from the Sun than Jupiter's average distance from the Sun?

   ____

8. How many AU shorter is Earth's average distance from the Sun than Saturn's average distance from the Sun?

   ____

9. How many AU is the average distance from the sun to the planet Uranus?

   A 5 **C** 19
   B 10 **D** 30

10. How many AU is the average distance from the Sun to Neptune?

    F 5 **H** 19
    G 10 **J** 30

Name____

Make Bar and Double-Bar Graphs

Use the data in the table to make two bar graphs. Then make a double-bar graph. Use the space provided below.

1.

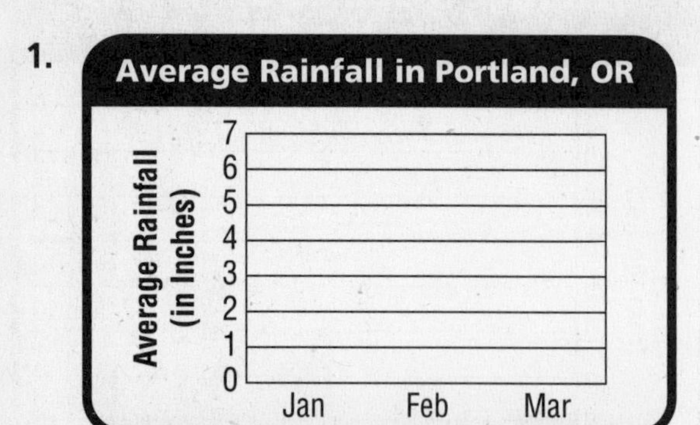

Average Rainfall Boulder, CO

Average Rainfall (in Inches)

7, 6, 5, 4, 3, 2, 1, 0

Jan Feb Mar

Average Rainfall (in inches)			
City	**Jan**	**Feb**	**Mar**
Portland, OR	6	5	5
Boulder, CO	1	1	2

2.

Average Rainfall (in inches)

7, 6, 5, 4, 3, 2, 1, 0

Boulder, CO
Portland, OR

Jan Feb Mar
Month

For 3–6, use the graphs you made.

3. Which city gets the most rainfall from January through March?

____

4. During which month does Boulder get the most rainfall?

____

5. Which city has a greater range of inches of rainfall in the three months?

____

6. Compare the two cities. During which month is the difference in rainfall the greatest? How great?

____

For 7–8, use the Favorite Sports graph at the right.

7. What is the range of the data?

____

8. How many more girls than boys like soccer the most?

____

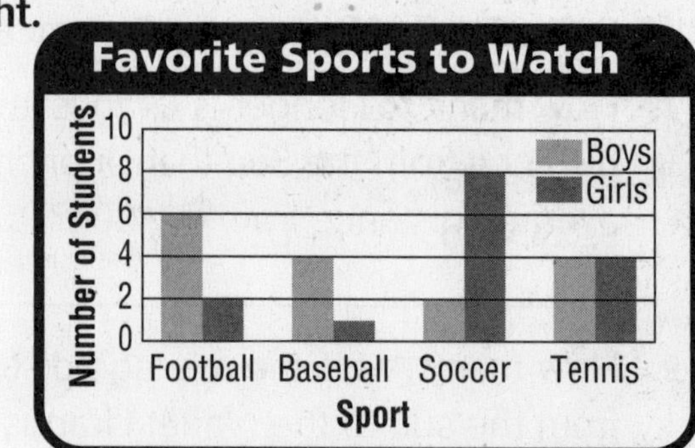

Name________________________________

Lesson 19.7

Problem Solving Workshop Skill: Make Generalizations

Problem Solving Skill Practice

USE DATA For 1–3, use the weight range chart. Make a generalization. Then solve the problem.

1. Complete the chart.

Known Information

The chart shows the healthy weight range for adults. Adults who are ■ tall should weight between ■ and 184 pounds. Healthy adults who weigh between 139 and 174 pounds may be about ■ tall. An adult who is 5′7″ should weigh between ■ and ■.

- Minimum weights increase in ■-pound increments.
- Maximum weights increase in ■-pound increments.

Height (ft, in.)	Adult Weight Ranges (in pounds)	
	Minimum	Maximum
5'7"	127	159
5'8"	131	164
5'9"	135	169
5'10"	139	174
5'11"	143	179
6'0"	147	184

2. Kosi is 5′9″ tall. What is a healthy weight range for Kosi?

____

3. Gwen is a healthy adult who weighs 135 pounds. According to the chart, what might be Gwen's range in height?

____

Mixed Applications

For 4–7, use the weight range chart.

4. How much greater is the weight range of a healthy adult who is 6′0″ tall than one who is 5′7″?

____

5. Gino weighs 180 pounds. About how much more does Gino weigh than Tu who is at maximum weight for 5′9″?

____

6. If the pattern continues, what will be the range of healthy weights for an adult who is 6′1″ tall?

____

7. **Pose a Problem** Look at Exercise 4. Change the numbers to make a new problem.

____

Name____

Algebra: Graph Ordered Pairs

For 1–4, use the grid at the right. Write the ordered pair for each point.

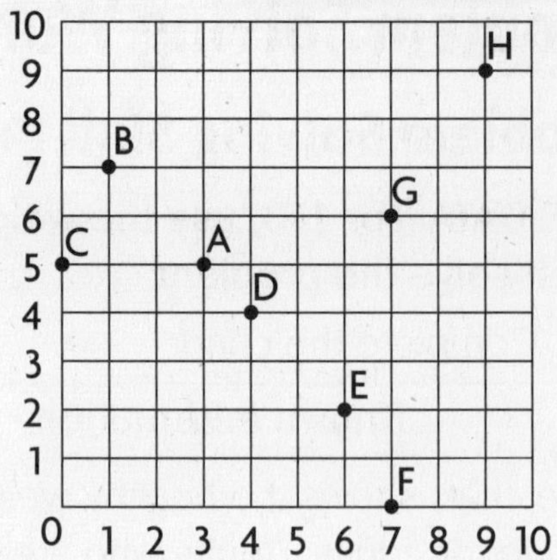

1. C (■, ■)

2. G (■, ■)

3. D (■, ■)

4. B (■, ■)

For 5–6, write the ordered pairs for each table. Then use the grid on the right to graph the ordered pairs.

5.

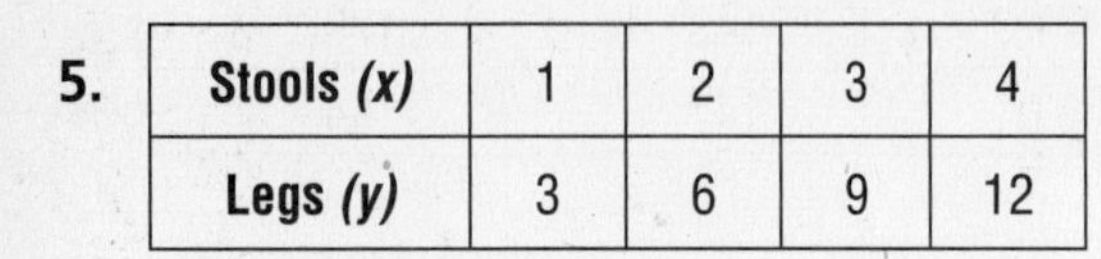

Stools (x)	1	2	3	4
Legs (y)	3	6	9	12

(■, ■), (■, ■), (■, ■), (■, ■)

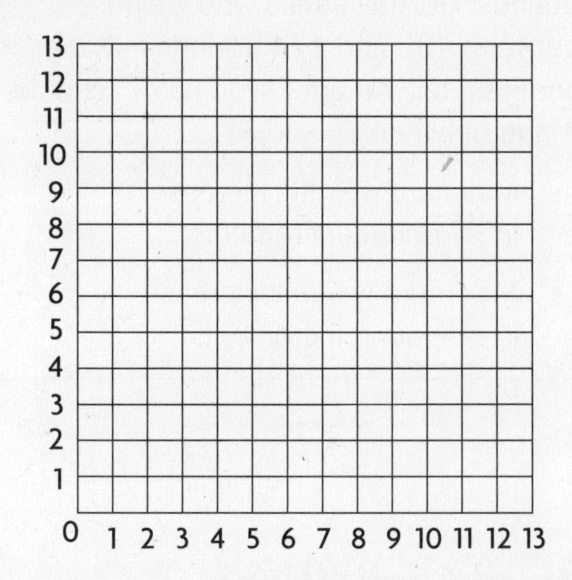

6.

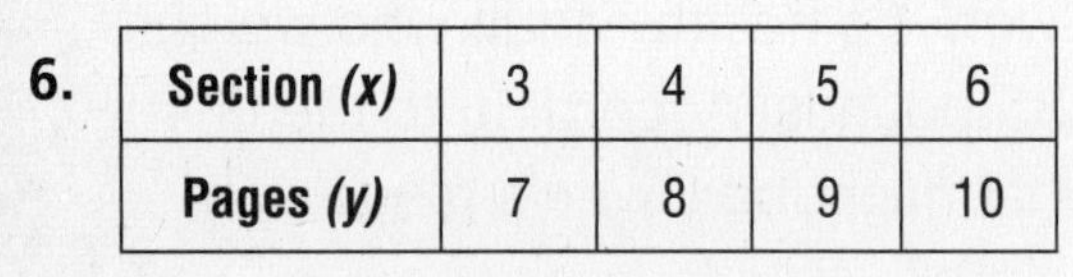

Section (x)	3	4	5	6
Pages (y)	7	8	9	10

(■, ■), (■, ■), (■, ■), (■, ■)

Problem Solving and TAKS Prep

7. Look at Exercise 6. Fabio is making a book in which the sections have increasing numbers of pages. How many pages will section 10 have?

____

8. Look at Exercise 5. Kip is making three-legged stools. If he has enough seats to make stools using 24 legs, how many stools can he make?

____

9. Use the coordinate grid at the top of the page. What is the ordered pair for point F?

A (6, 2) **C** (3, 5)

B (9, 9) **D** (7, 0)

10. Marty uses 4 cups of flour to make a cake. How many cakes can Marty make with 28 cups of flour?

F 4 **H** 7

G 6 **J** 8

Name____ **Lesson 20.1**

Model Combinations

Write the letters of the word PENCIL on a strip of paper with 6 sections. Use the paper to model combinations of the given number of letters.

1. 1 letter

____

2. 6 letters

____

3. 3 letters

____

4. 4 letters

____

5. 5 letters

____

Make a table in the space provided to solve.

6. How many combinations of 3 names can be made from the names David, Sarah, Daphne, and Trent?

David	Sarah	Daphne	Trent

____

7. How many combinations of 2 letters can be made from the letters in the word WORK?

W	O	R	K

____

8. How many combinations of 4 snacks can be made from apple, orange, pear, carrot, celery, and snap pea?

Apple	Orange	Pear	Carrot	Celery	Snap Pea

____

Name

Combinations

Make a model to find all the possible combinations.

1. Dinner choices
   sauce: marinara, cheese, meat
   pasta: spaghetti, linguine, macaroni

2. Snack choices
   fruit: apple, orange, plum
   vegetable: celery, carrot

3. Writing material choices
   pen: ballpoint pen, fountain pen, flair pen, roller ball pen
   paper: white, red, yellow

4. Hamburger choices
   cheese: Swiss, cheddar, American, provolone
   spreads: ketchup, mustard, mayo

Problem Solving and TAKS Prep

5. In how many ways can the fruit in Exercise 2 be ordered?

6. In how many ways can the cheese in Exercise 4 be ordered?

USE DATA For 7–8, use the table.

7. How many different combinations of salads are possible?

Salad Choices	
Greens	**Dressings**
Mixed greens	Oil and vinegar
Romaine	Caesar
Red loose leaf	Ranch
	Honey mustard

A 7
B 9
C 12
D 16

8. Spinach is added as a fourth choice of greens to Exercise 9. List all of the possible combinations that now exist.

Name____

Tree Diagrams

Complete the tree diagram to find the number of possible combinations.

1. cats:
   breed: Angora, Siamese, Tabby
   colors: black, white, gray, cream

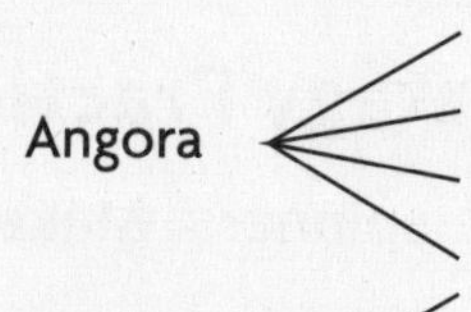

2. breakfast choices
   main: eggs, cereal, waffles, pancakes
   fruit: orange, apple, pear, berries

eggs

cereal

waffles

pancakes

Problem Solving and TAKS Prep

3. Gil has 4 different pairs of gloves and 5 different pairs of socks. How many different combinations of gloves and socks can Gil choose from?

   ____

4. **Pose a Problem** Look back at Exercise 2. Write a similar problem by changing the choices.

   ____

5. Tina is making doll clothes. She can make pants or a skirt using blue, red, or patterned fabric. How many different combinations are possible?

   A 3 **C** 9

   B 6 **D** 12

6. Eduardo is making a flag. He can choose between stars, stripes, or trees on a background of red, blue, or green. How many combinations do not use stripes or a red background?

   ____

Name____ Lesson 20.4

Problem Solving Workshop Strategy: Make an Organized List

Problem Solving Strategy Practice

USE DATA For 1–3, use the spinners. Make an organized list to solve.

1. Franco made these spinners for a school carnival game. What are the possible combinations?

____

____

____

____

2. To win, Gloria must spin both pointers for a total more than 6. Name the ways Gloria can win.

____

3. Patty can win if she spins both pointers for a total of more than 5. Name the ways Patty can win.

____

Mixed Strategy Practice

4. Pedro is making cards for a game. Each type of card will be a different color. The suits will be hearts and flags. In each suit, there will be 3 sets: numbers, letters, and symbols. How many colors will there be?

____

5. **Open Ended** You probably made an organized list to solve Exercise 4. What is another strategy you could use to solve it? Explain.

____

____

____

6. Jorge's father has driven his car 103,240 miles. His mother has driven hers 69,879. How much further has his father driven?

____

7. There are 110 students in fourth grade. Thirty-two take only music, 25 take only art, and 12 take both. How many students do not take art or music?

____

Name____ Lesson 20.5

More About Combinations

How many different pairs of items can there be if each pair must have 1 item from each category?

1. **Flower:** daisy, rose, pansy
   Leaf: elm, oak, laurel, pear

   ____

2. **Car:** sedan, convertible, SUV, van
   Color: red, blue, white, black, grey

   ____

3. **Cycles:** bicycle, tricycle, unicycle,
   Color: black, red, blue, green, yellow, white

   ____

4. **Timepieces:** wristwatch, pendant, pocket watch
   Time: Eastern, Central, Pacific

   ____

5. **Season:** summer, fall, winter, spring
   Weather: rainy, sunny, windy, cloudy

   ____

6. **Vehicles:** car, bus, truck, semi
   Year: 2006, 2007, 2008, 2009

   ____

Problem Solving and TAKS Prep

7. List the different ways that the hammer, the screwdriver and the hand drill can be arranged on a table.

   ____

   ____

   ____

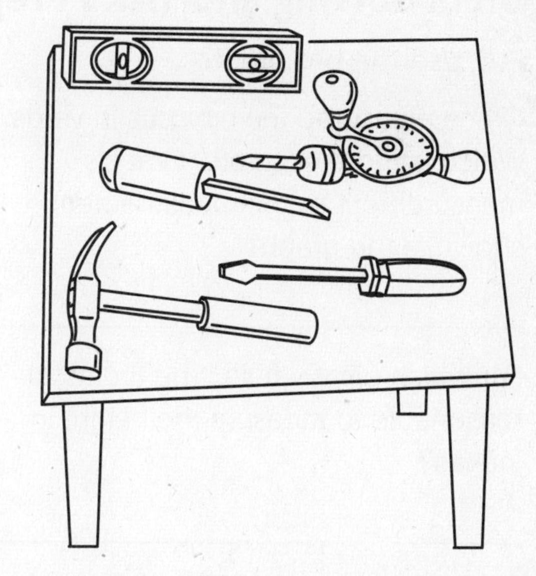

8. If every tool shown comes with a red or a black label, how many combinations of tools and labels are there?

   ____

9. Don has the four arrowheads shown below. In how many different ways can Don mount the four arrowheads on a display?

   A 24 **C** 12
   B 18 **D** 9

10. Jan has 5 charms to put on her bracelet. In how many different ways can Jan attach the five charms?

   F 100 **H** 125
   G 120 **J** 150

Name____

Measure Fractional Parts

Estimate to the nearest $\frac{1}{2}$ inch. Then measure to the nearest $\frac{1}{8}$ inch.

1.

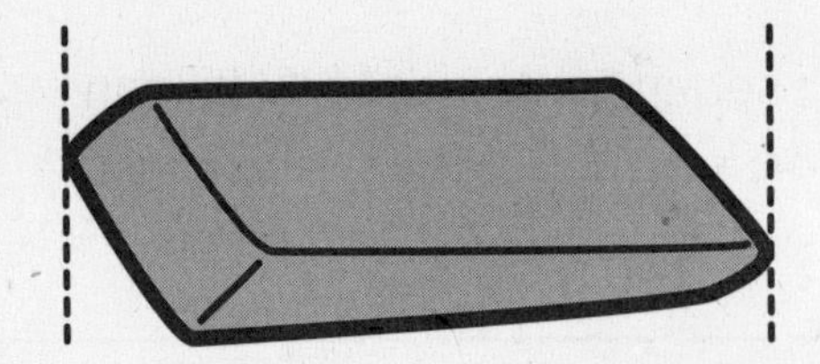

____

2.

____

3.

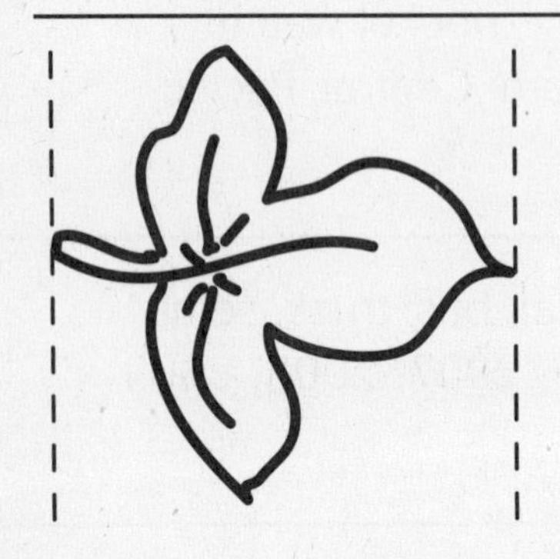

____

4.

____

Problem Solving and TAKS Prep

For 5–6, use the bar graph.

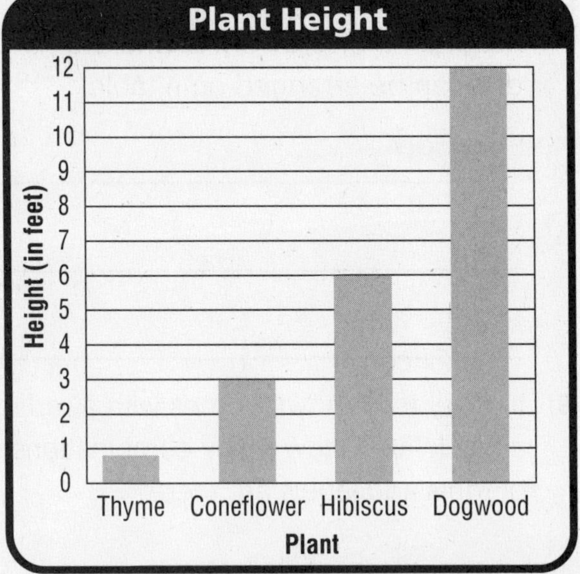

5. Grace measured the hibiscus in yards. Which is more accurate, Grace's measurement or the measurement shown in the graph?

____

6. For which plant would it be the most reasonable to measure the height in inches?

____

7. What is the length of the yarn to the nearest $\frac{1}{8}$ inch?

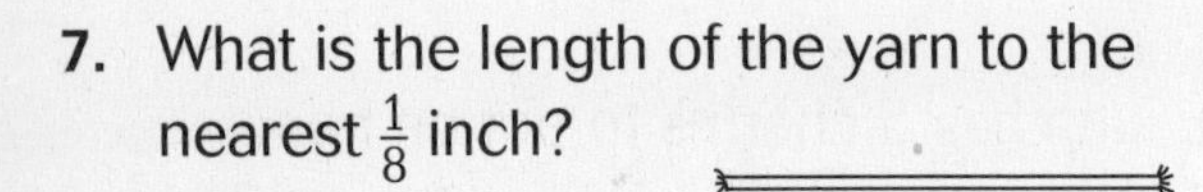

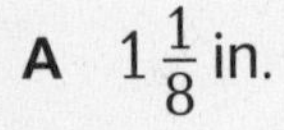

A $1\frac{1}{8}$ in. **C** $1\frac{3}{8}$ in.

B $1\frac{1}{4}$ in. **D** $1\frac{1}{2}$ in.

8. Haley painted a picture using paintbrushes that were $1\frac{1}{8}$ in wide, $1\frac{1}{4}$ in. wide, $1\frac{3}{8}$ in. wide, and $1\frac{1}{2}$ in. wide. Order the width of the paintbrushes from greatest to least.

____

Name________________________________ Lesson 21.2

Algebra: Change Customary Linear Units

Complete. Tell whether you multiply or divide. Circle M for multiply, D for divide.

M or D
1. 7 yd = ■ in.

M or D
2. ■ miles = 7,040 yd

M or D
3. 15 ft = ■ yd

M or D
4. ■ in. = 7 ft

M or D
5. 144 in. = ■ ft

M or D
6. ■ yd = 288 in.

Compare. Write <, >, or = in each ●.

7. 60 in. ● 5 ft
8. 36 ft ● 11 yd
9. 98 in. ● 5 yd
10. 3,520 yd ● 1 mi

Write an equation you can use to complete each table. Then complete the table.

11.

Inches, n	48	60	72	84	96
Feet, f	4	5	■	■	■

____

12.

Miles, m	1	2	3	4	5
Yards, Y	1,760	3,520	■	■	■

____

Problem Solving and TAKS Prep

For 13–14, use the Fabric Requirements table.

13. Fahra has 8 yards of fabric. How much will she have left over if she makes 3 small costumes?

____

14. How many inches of fabric are needed to make a costume for a large dog?

____

Fabric Requirements	
Dog Costume	
Size	Yards
Small	$2\frac{5}{8}$
Medium	$2\frac{7}{8}$
Large	$3\frac{1}{8}$

15. George is 75 inches tall. How many feet tall is he?

____

16. Barbara is 64 inches tall. How many feet tall is she?

____

Name________________________________

Weight

Complete. Tell whether you multiply or divide.
Circle M for multiply, D for divide.

M or D
1. 176 oz = ■ lb

M or D
2. ■ oz = 5 lb

M or D
3. 7 T = ■ lb

M or D
4. ■ oz = 12 lb

M or D
5. 320 oz = ■ lb

M or D
6. ■ T = 8,000 lb

Choose the more reasonable measurement.

7. 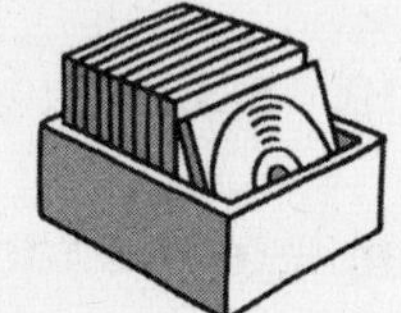

4 oz or 4 lb

8. 

6 oz or 6 lb

9. 

about 16 lb or 16 T

10. 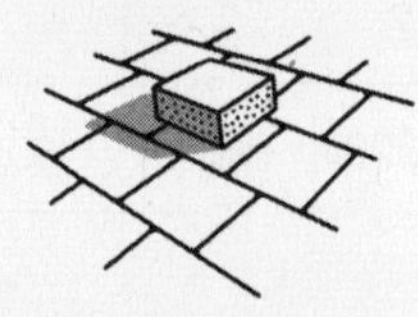

about 7 lb or 7 T

11. 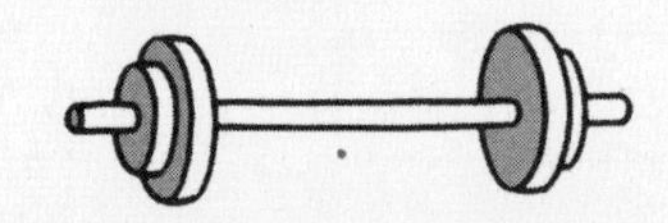

100 oz or 100 lb

12. 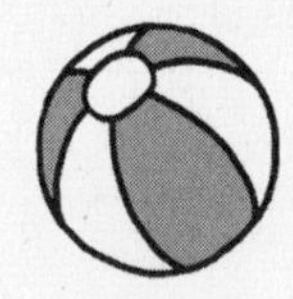

5 oz or 5 lb

Problem Solving and TAKS Prep

USE DATA For 13–14, use the table.

13. In pounds and ounces, how much corn and meat does a crow eat in a week?

____

14. How many more pounds of food does a tiger eat in 2 days than a crow eats 32 days?

____

Adult Animal Daily Food Consumption		
Animal	**Diet**	**Amount**
Crow	Corn, meat	11 oz
Tiger	Meat	14 lb
Panda	Bamboo	26 lb

15. Cara bought five 4-lb bags of cat food. How many ounces is that?

A 64 ounces **C** 20 ounces
B 320 ounces **D** 160 ounces

16. Two Goliath toads can weigh up to 208 ounces. How many pounds is that?

F 9 lbs **H** 13 lbs
G 11 lbs **J** 15 lbs

Name________________________________ Lesson 21.4

Customary Capacity

Complete each table. Change the units.

1.

quarts, q	5	10	15
cups, c			

2.

pints, p		14	16
quarts, q	3		

3.

gallons, g		4	
pints, p	16		48

4.

tsp			
tbsp	5	10	15

5.

cups, c		20	
pints, p	8		12

6.

gallons, g			
quarts, q	16	24	36

Circle the more reasonable unit of capacity.

7. 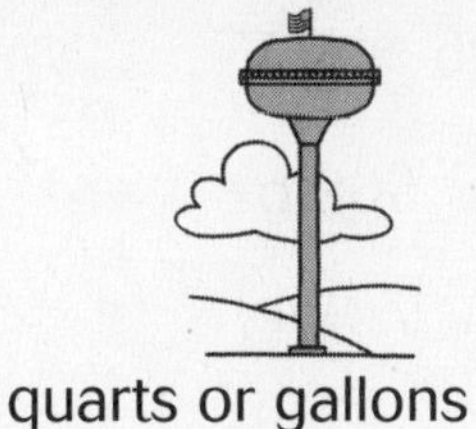
quarts or gallons

8. 
teaspoons or pints

9. 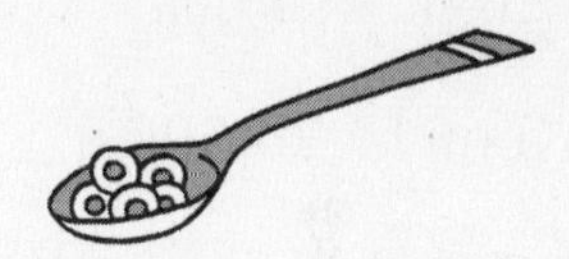
tablespoons or quarts

10. 
teaspoons or cups

11. 
teaspoons or quarts

12. 
pints or gallons

Problem Solving and TAKS Prep

USE DATA For 13–15, use the recipe.

Green Tea Punch

3 pints green tea

$\frac{1}{2}$ pint orange drink

1 cup lemon juice

1 quart lemonade

Serves 12

13. How many total fluid ounces are in the recipe?

____

14. How many ounces are in one serving of the tea punch?

____

15. Sandy makes 4 gallons of Green Tea Punch for a meeting. How many pints of Green Tea Punch does she make?

____

16. Randy filled 48 pint jars with jelly made from the grapes in his garden. How many gallons of jelly did he make? Explain.

____

Name____

Algebra: Change Units

Complete. Tell whether you multiply or divide. Circle M or D.

1. M or D

   96 oz = ■ lb

2. M or D

   ■ c = 104 pt

3. M or D

   9 yd = ■ in

4. M or D

   7 qt = ■ c

5. M or D

   512 oz = ■ lb

6. M or D

   ■ lb = 14 T

ALGEBRA Compare. Write >, <, or = for each ●.

7. 30 pt ● 4 gal
8. 7 lb ● 112 oz
9. 6,500 lb ● 3 T
10. 5 mi ● 26,500 ft
11. 398 in ● 11 yd
12. 35 lb ● 560 oz

Complete the table. Change the units.

13.

Cups, c	■	■	16
Quarts, q	8	6	■

14.

Gallons, g	12	■	16
Pints, p	■	112	■

15.

Inches, i	144	288	■
Yards, y	■	■	18

Problem Solving and TAKS Prep

16. Georgia has 200 inches of fabric to sew a pattern for a costume that calls for 6 yards. Does she have enough fabric to complete make the pattern?

    ____

17. Benny wants to ride his bike more laps than anyone else. The track is 1,100 yards around. If Benny completes 80 rounds, how many feet will he have traveled around the track?

    ____

18. Uma uses 5 pints of grape, 3 pints of lemon, and 7 quarts of orange juice to make a punch. How many cups of punch does she make in all?

    ____

19. An eagle searching for prey circled and then flew straight north for a distance of 3 miles. How many feet did it travel?

    ____

Name

Lesson 21.6

Problem Solving Workshop Strategy: Compare Strategies

Problem Solving Strategy Practice

Choose a strategy to solve. Explain your choice.

1. Karen visited an aquarium with a tank for small fish that was 8 feet long. How many inches long is the tank?

2. Bea fried 36 4-ounce servings of catfish. How many pounds of fish did Bea fry in all?

3. Lyle made 7 quarts of tartar sauce for a club outing. How many $\frac{1}{2}$-pint jars will Lyle need to store all the tartar sauce?

4. Mitch gave a mile of fishing line to each of 4 competitors in a fishing tournament. How many feet of fishing line did Mitch give away?

Problem Solving and Test Prep

USE DATA For 5–8, use the table.

5. If the next largest Searobin caught weighed 80 ounces, how many ounces did the two largest Searobins weigh together?

6. Laid end-to-end, how long a line in inches would all the biggest fish make?

Scout Deep Sea Fishing Trip Largest Fish Caught

Fish	Weight (lb)	Length (ft)
Cod	85	6
Flounder	15	2
Striped bass	45	4
Bluefish	12	3
Searobin	7	$1\frac{1}{4}$

7. How many more ounces did the flounder weigh than the bluefish?

8. Open Ended What strategy would you use to find out the weight in ounces of the entire catch? Explain your choice.

Name____________________ Lesson 22.1

Metric Length

Choose the most reasonable unit of measure. Write *mm, cm, dm, m,* or *km.*

1.  ____________

2.  ____________

3.  ____________

Estimate the nearest centimeter. Then measure to the nearest half centimeter. Write your answer as a decimal.

4. 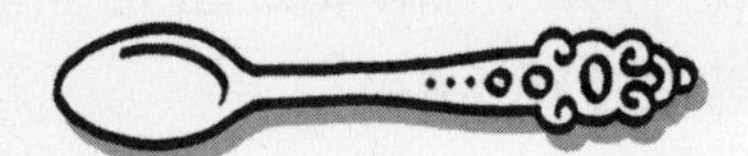 ____________

5.  ____________

Estimate to the nearest half centimeter. Then measure to the nearest millimeter.

6.  ____________

7.  ____________

Problem Solving and TAKS Prep

8. In Exercise 1, what is the length to the nearest millimeter of the ant?

____

9. What is the most reasonable unit of measure that can be used to measure the length of this workbook?

A millimeters **C** meters

B centimeters **D** kilometers

Name____ Lesson 22.2

Mass

Choose the more reasonable measurement.

1. 

20 g or 20 kg

2. 

14,500 kg or 14,500 g

3. 

5,220 g or 1.22 kg

4. 

8,000 g or 1 kg

5. 

300 g or 300 kg

6. 

23 g or 23 kg

Compare the mass of each object to a kilogram. Write *about 1 kilogram, less than 1 kilogram,* or *more than 1 kilogram.*

7. 

____

8. 

____

9. 

____

Problem Solving and TAKS Prep

Mass of Sports Balls

Basket ball
616 grams

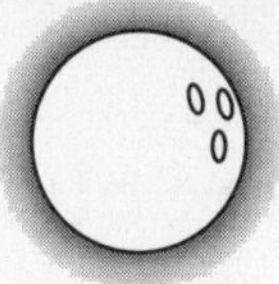

Bowling ball
6 kilograms

Table tennis ball
2.5 grams

Tennis ball
57 grams

USE DATA For 10–11, use the table.

10. What is the total mass in grams of one of each of all the sports balls?

____

11. Which has the greater mass, 1 basketball or 11 tennis balls?

____

12. Wanda needs a kilogram of peanut butter. How many 510-gram jars will she need to buy?

A 2 **C** 4

B 3 **D** 5

13. Ed bought a 1.02-kg box of ramen noodles. There are 12 packs in each box. What is the mass of each pack?

F 0.85 **H** 85

G 8.5 **J** 0.085

Name______________________________

Capacity

Choose the most reasonable measurement. Circle *a*, *b*, or *c*.

1. a. 8 L
b. 8 mL
c. 20 mL

2. a. 4 L
b. 1 L
c. 10 mL

3. a. 1 L
b. 3 mL
c. 3 L

Estimate and tell whether each object had a capacity of *about a liter, less than a liter,* or *more than a liter.*

4.

____

5.

____

6.

____

Compare. Write >, <, or = in each ●.

7. 25 L ● 52 L

8. 1,300 mL ● 3,100 mL

9. 9 L ● 9,000 mL

Problem Solving and TAKS Prep

10. Kyle uses 1 liter of gas to drive 12 km. How far can he drive with 500 mL of gas?

____

11. Patti uses 2 L of sauce for every 4 servings of spaghetti. How many liters does she need to make 12 servings?

____

12. Jane is measuring a container's capacity in liters. Which of the following objects is she most likely measuring?

A gas tank
B tea cup
C the ocean
D perfume bottle

13. Hal is measuring a container's capacity in mL. Which of the following objects is he most likely measuring?

F eye dropper
G gas tank
H the ocean
J swimming pool

Name________________________________ Lesson 22.4

Problem Solving Workshop Strategy: Make a Table

Problem Solving Strategy Practice

Make a table to solve each problem.

1. Sam wants to figure out how tall his 2-year-old brother will be as an adult. He went to the library and found this information: a boy who is 87 cm at age two will be about 174 cm as an adult, a boy who is 90 cm at age two will be about 180 cm as an adult, and a boy who is about 92 cm at age two will be about 184 cm as an adult. Sam knows that his brother is 91 cm tall now. About how tall will Sam's brother be as an adult?

____

2. A pattern of mosaic tiles is lined up in this order: 8 cm, 5 cm, 8 cm, and 12 cm. How many meters long will be pattern be if it repeats 50 times?

____

Mixed Strategy Practice

USE DATA For 3–4, use the picture.

History's Tallest and Shortest People

Person	Height
Tallest man	2.720 m
Shortest man	57 cm
Tallest woman	2.48 m
Shortest woman	61 cm

3. Draw a table to show the heights of History's Tallest and Shortest People in order from greatest height to least height. Include your height in the table.

4. How many decimeters difference is there between the height of the tallest woman and that of the shortest man?

____

5. Dick is buying a plant that will grow 10.4 cm in 2 days. How many millimeters will it grow per day?

____

Name________________________________

Choose the Appropriate Tool and Unit

Choose the tool and unit to measure each.

Tools	Units
ruler	mi
measuring cup	g
spring scale	in.
yardstick	mm
meterstick	L
dropper	mL
odometer	kg

1. width of a window ____________
2. capacity of a bottle ____________
3. weight of an apple ____________
4. distance to the theater ____________
5. mass of a raindrop ____________
6. length of a staple ____________
7. mass of a dozen roses ____________
8. capacity of a tea cup ____________
9. length of a driveway ____________
10. weight of a lamp ____________
11. length of a fingernail ____________
12. capacity of a punchbowl ____________

Problem Solving and TAKS Prep

13. Which customary and metric tools would you use to measure the length of a garden path?

____

14. Which metric and customary units would you use to measure the distance from the cafeteria to the art room in your school?

____

15. What tool would you use to measure the capacity of a container of milk?

A dropper
B yardstick
C measuring cup
D spring scale

16. Which customary and metric units would you use to measure the height of a countertop?

____

Name____ Lesson 22.6

Reasonable Estimates

Draw a circle around the more reasonable estimate.

1.

2 cm or 2 m

2.

12,000 g or 12,000 kg

3.

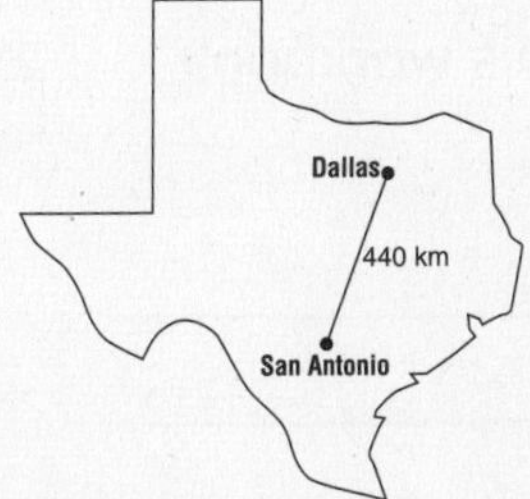

275 mi or 275 yards

4.

5 L or 5 mL

5.

4 oz or 4 lb

6.

3 qt or 3 gal

Problem Solving and Taks Prep

For 7–8, use the tall tale.

Harry's Texas Tall Tale

Everything is bigger in Texas. Harry traveled 52,800 feet to the store in a neighboring town to buy a pair of boots. The boots he bought weighed 5 tons. When he stopped for lunch, he drank an 80 liter glass of milk.

7. What is a reasonable metric estimate of the distance to a neighboring town?

____

8. What is a reasonable customary estimate of how much a pair of boots should weigh?

____

9. Diana bought cheese to make sandwiches for her four friends. Which is the most reasonable estimate of the amount of cheese she bought?

A 1 ton **C** 5 pounds

B 500 pounds **D** 1 pound

10. Qixo likes watching the black bears at the zoo. Which is the better estimate of how much a black bear weighs, 200 kg or 2,000 kg?

____

Name____ Lesson 23.1

Estimate and Measure Perimeter

Use string to estimate and measure the perimeter of each object.

1. this workbook
2. the doorway to your bedroom
3. the face of a TV
4. the door of your refrigerator

Find the perimeter of each figure.

5. ____________
6. ____________
7. ____________
8. ____________
9. ____________

Problem Solving and TAKS Prep

For 10–11, use the dot paper above.

10. Draw and label a square with a perimeter of 8 units. What are the lengths of the sides?

____

11. Draw and label a square with a perimeter of 16 units. What are the lengths of the sides?

____

12. Which rectangle has the greatest perimeter?

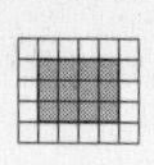

C 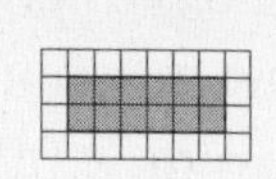

B 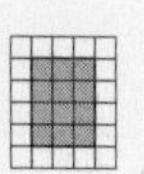

D 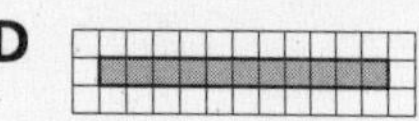

13. Which rectangle has the greatest perimeter?

F 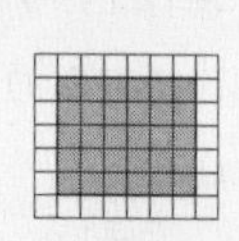

H 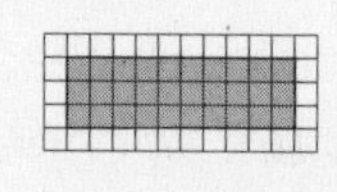

G 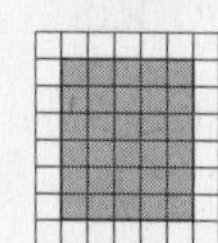

J 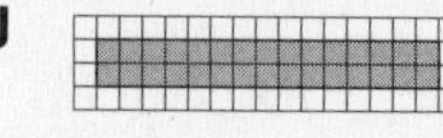

Name____

Perimeter

Find the perimeter.

1. 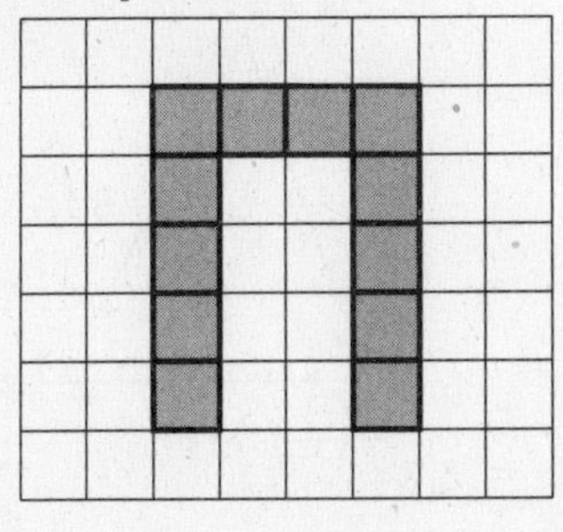

2. 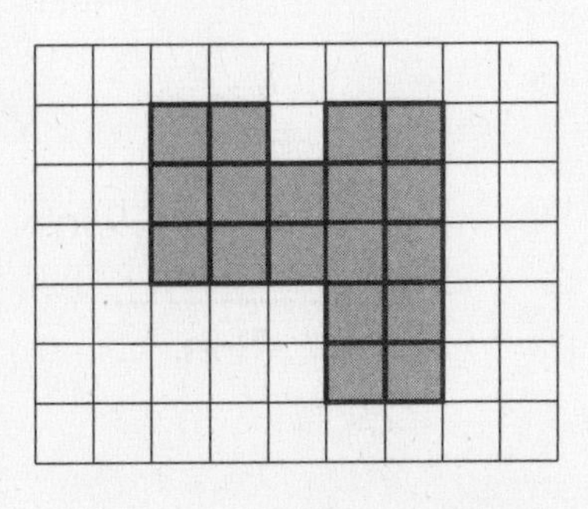

3. 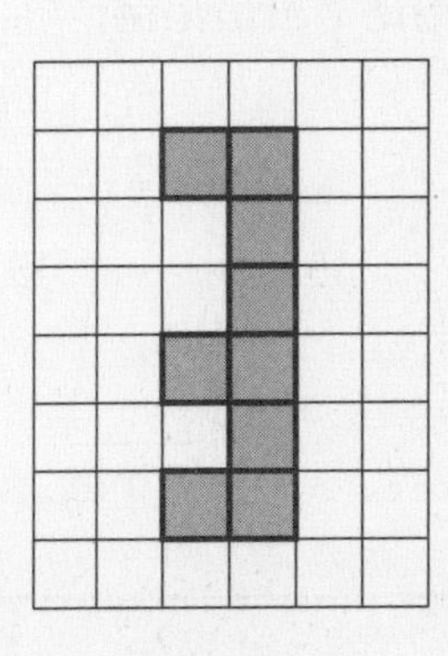

4. 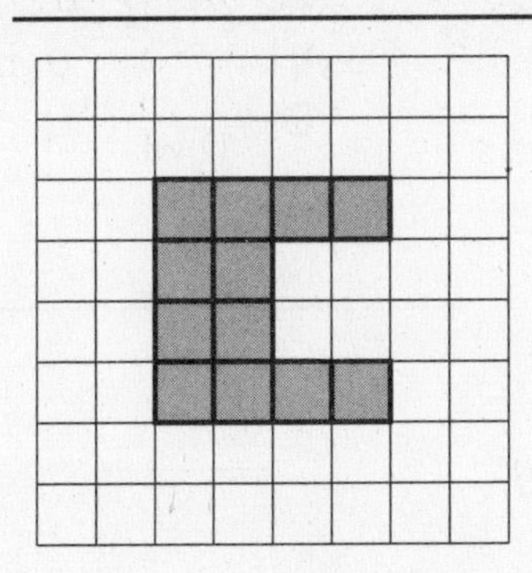

5. 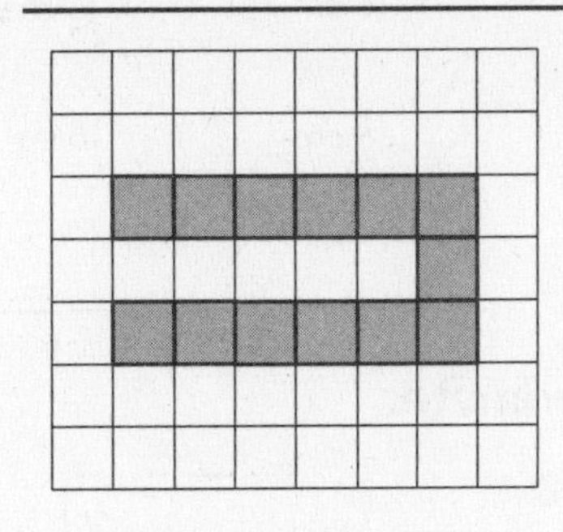

6. 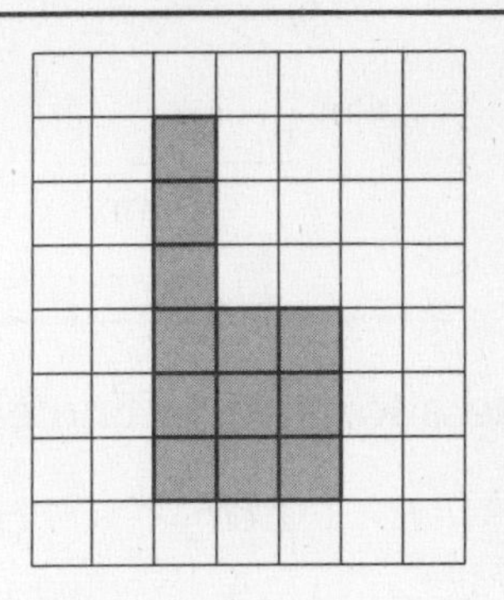

Problem Solving and TAKS Prep

USE DATA For 7–8, use the map.

7. Estimate the perimeter shown by the straight black lines.?

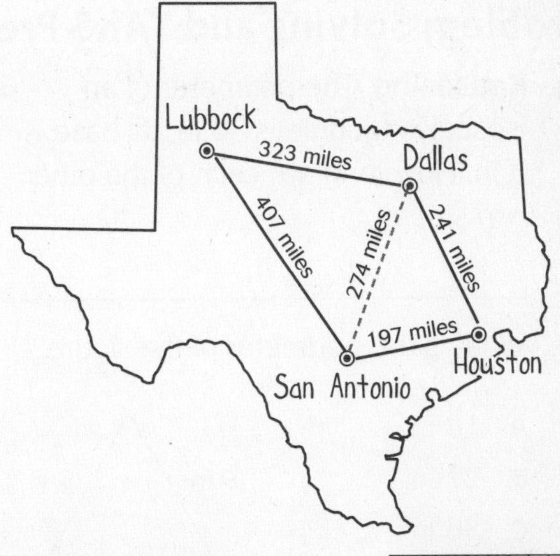

8. About how much greater is the perimeter shown by the solid black lines than the distance from San Antonio to Houston to Dallas?

9. What is the perimeter of this figure?

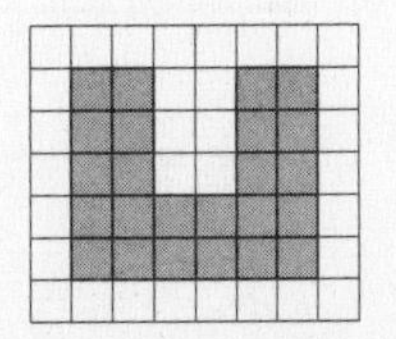

10. What is the perimeter of this figure?

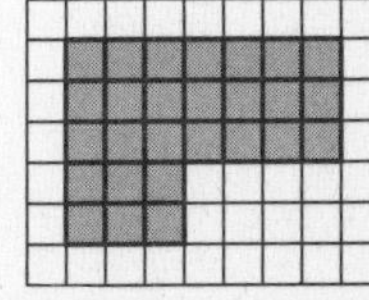

Name____ **Lesson 23.3**

Algebra: Find Perimeter

Find the perimeter.

1.

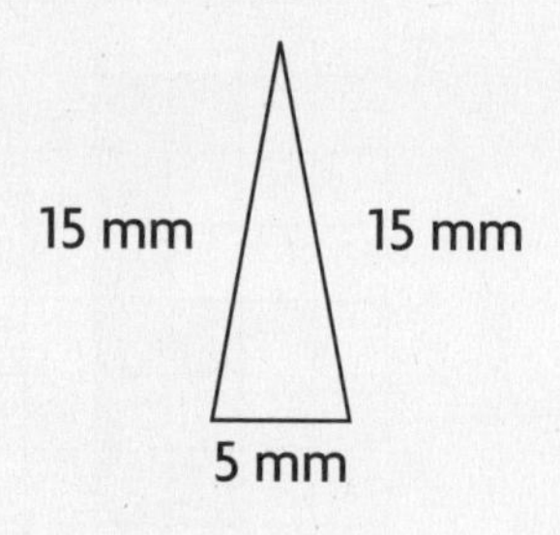

____

2.

7 in.
4 in.
9 in.
5 in.
11 in.

____

3.

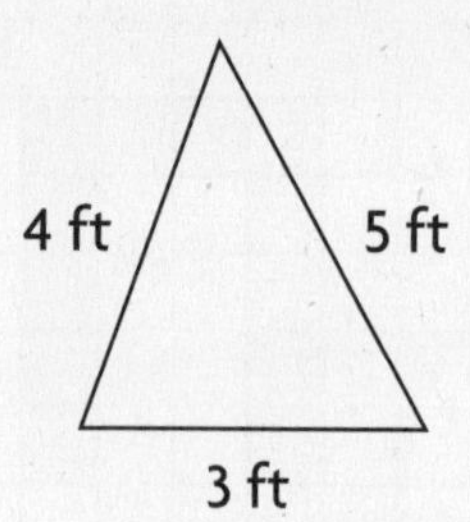

____

4.

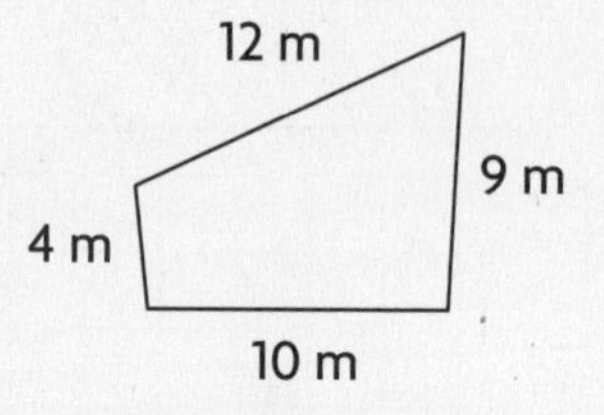

____

5.

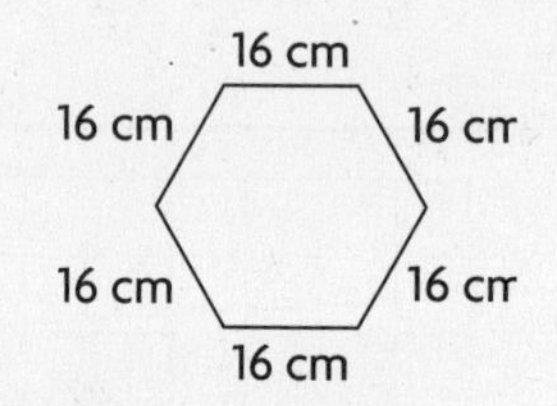

____

6.

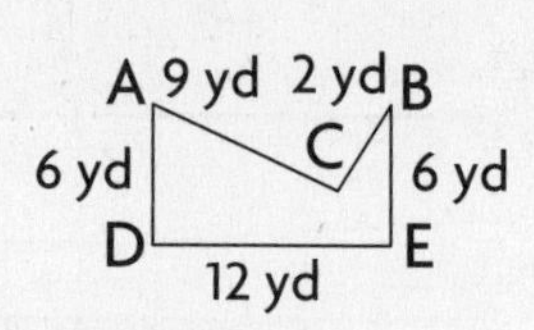

____

Use a formula to find each perimeter.

7.

7 cm
7 cm
7 cm
7 cm
7 cm

____

8.

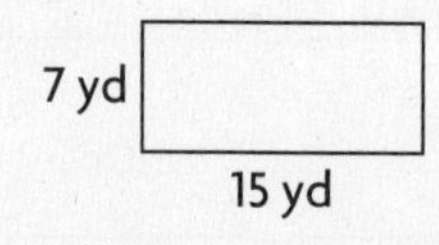

____

9.

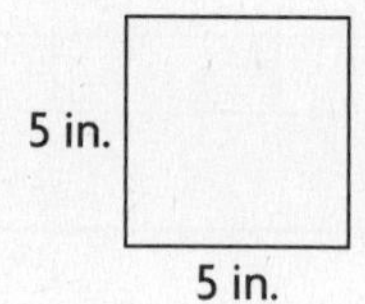

____

Problem Solving and TAKS Prep

10. Reasoning The perimeter of an isosceles triangle is 30 in. Its base is 8 in. How long are each of the other two sides?

____

11. Reasoning The perimeter of a rectangle is 46 ft. The width is 10 ft. What is the length?

____

12. What is the perimeter of this figure?

A 18 in.

B 27 in.

C 36 in.

D 45 in.

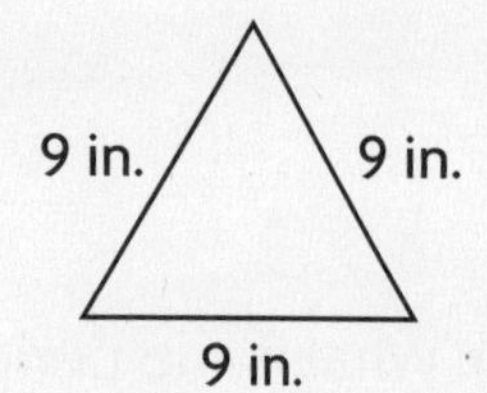

13. What is the perimeter of an equilateral hexagon with sides 6 cm long? Explain.

____

____

Name___________________________________ Lesson 23.4

Problem Solving Workshop Strategy: Compare Strategies

Problem Solving Strategy Practice

1. Lee is making a brick border around her square garden. Each side is 15 feet long. Each brick is a foot long. Will 54 bricks be enough? If not, how many more does she need?

____

2. Devon wants to paste a braided border around the edge of a picture frame that is 10 by 12. Then he wants to paste a fringe around the edge of the braided border. The braid is 2 inches wide. How much braid and fringe will he need?

____

3. Leon is sewing a beaded border around a rectangular blanket with sides are 1 yd by 2 yd. How many feet of beads does he need?

____

4. Joan is pasting 1 jewel per inch on each side of 4 squares with sides of 3, 4, 5, and 6 inches. How many jewels does she need?

____

Problem Solving and TAKS Prep

Choose a strategy to solve. Explain your choice.

USE DATA For 5–6, use the table.

5. Grant bought a 5-by-8 piece of carpeting and 9 yards of fencing. How much did he spend?

____

Bart's Building Supplies	
Supply	**Cost**
How-to Book	$15
Outdoor Carpet	$8/square foot
Fencing	$15/yard

6. Mr. Daley spent $195 for 3 how-to books and some fencing. How many yards of fencing did he buy?

____

7. The perimeter of a triangle is 27 inches. The sides are equal. What is the length of each side?

____

Name____________________

Estimate Area

Estimate the area of each figure. Each unit stands for 1 sq m.

1.

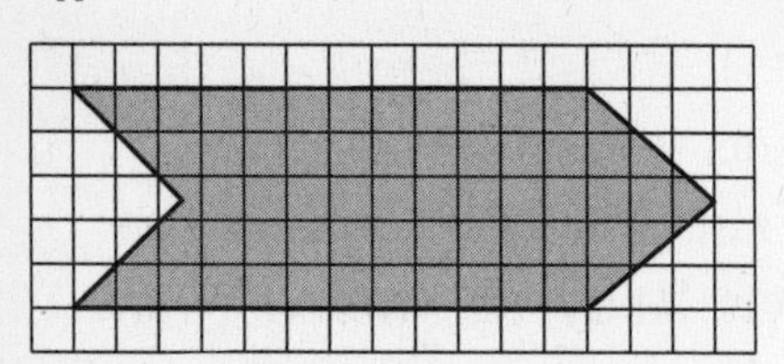

2.

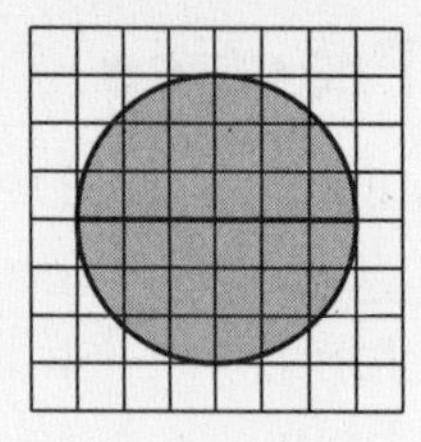

3.

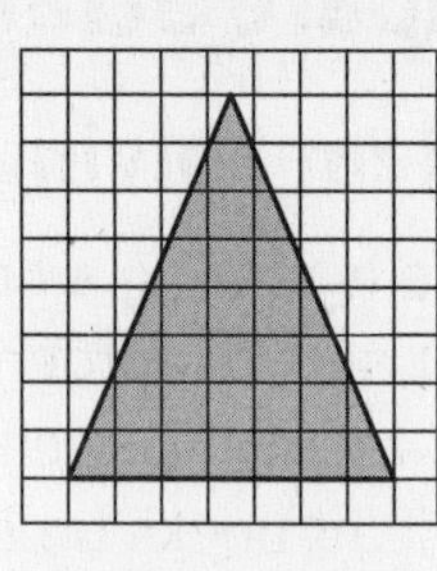

____________ ____________ ____________

4.

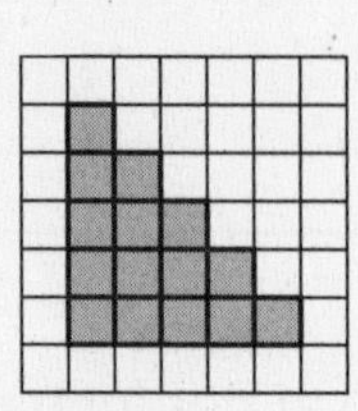

5.

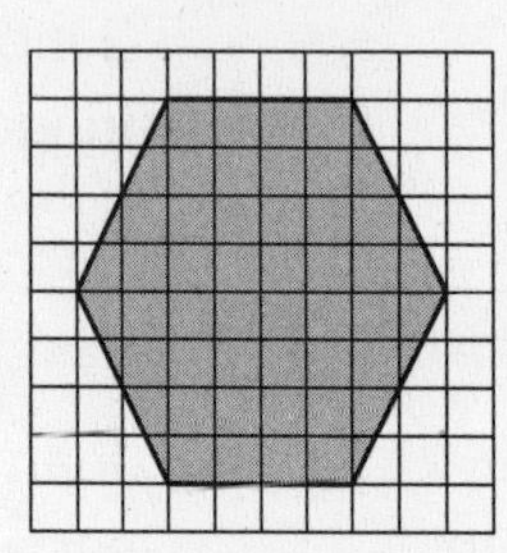

6.

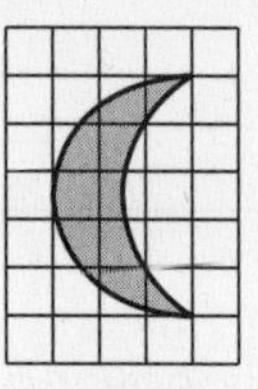

____________ ____________ ____________

Draw each figure on the grid paper at the right. Then estimate the areas in square units.

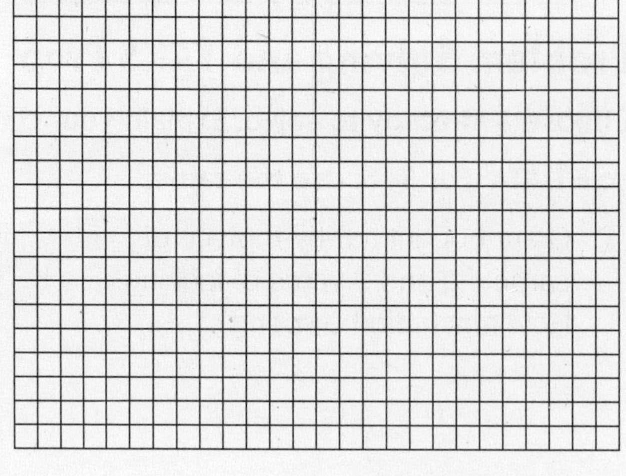

7. hexagon

8. right triangle

9. figure with straight lines

10. figure with curved and straight lines

USE DATA For 11–12, use the diagram.

Floor Plan

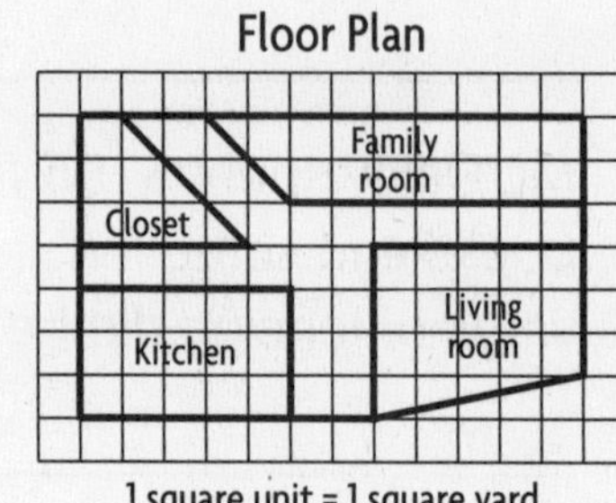

1 square unit = 1 square yard

11. About how many square yards is the hallway?

____

12. About how many square yards is the closet?

____

Name____________________ **Lesson 23.6**

Algebra: Find Area

Find the area.

1.

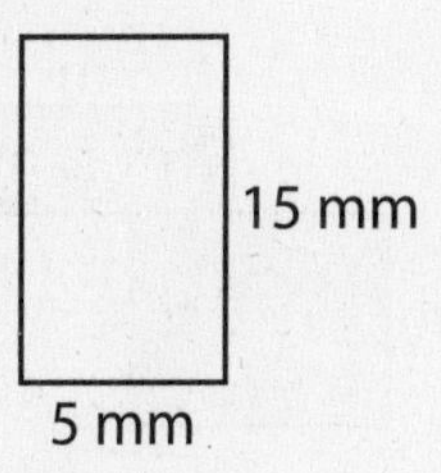

____

2.

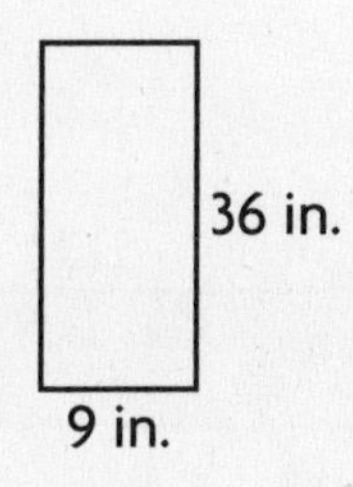

____

3.

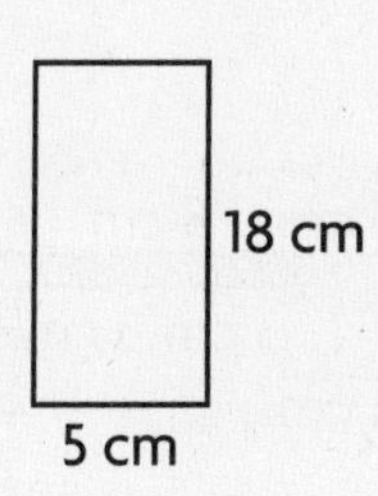

____

Use a centimeter ruler to measure each figure.

Find the area and perimeter.

4.

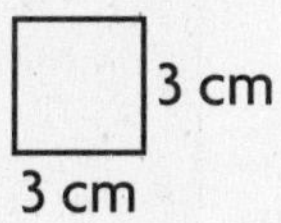

____

____

5.

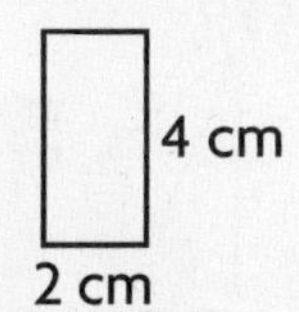

____

____

6.

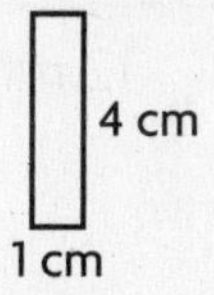

____

____

Problem Solving and TAKS Prep

For 7–8, use the diagram.

45 ft
Lawn
30 ft
30 ft
7 ft *Patio*
8 ft *Patio*
15 ft

7. What is the area and perimeter of the entire patio?

____

8. How much smaller is the area of the patio than the area of the lawn?

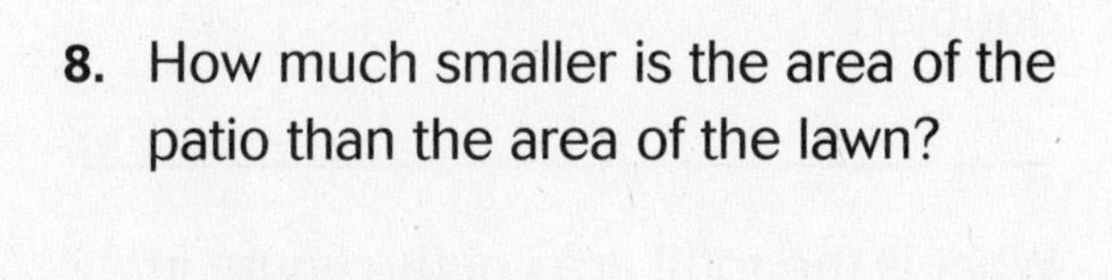

____

9. What is the area of this figure?

A 152 sq ft

B 162 sq ft

C 180 sq ft

D 200 sq ft

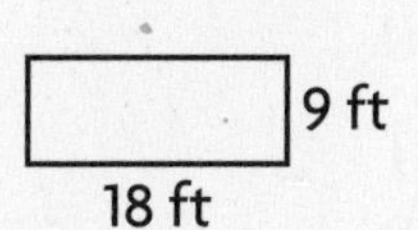

10. Use a formula to find the area of a rectangle that is 7 cm by 35 cm.

____

Name______________________________ **Lesson 23.7**

Perimeter and Area of Complex Figures

Find the perimeter and area of each figure

1.

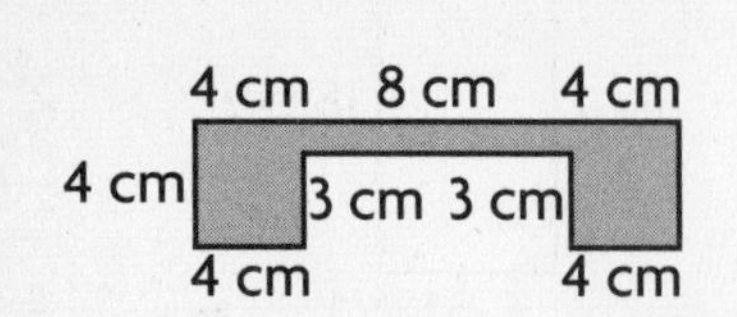

____

____

2.

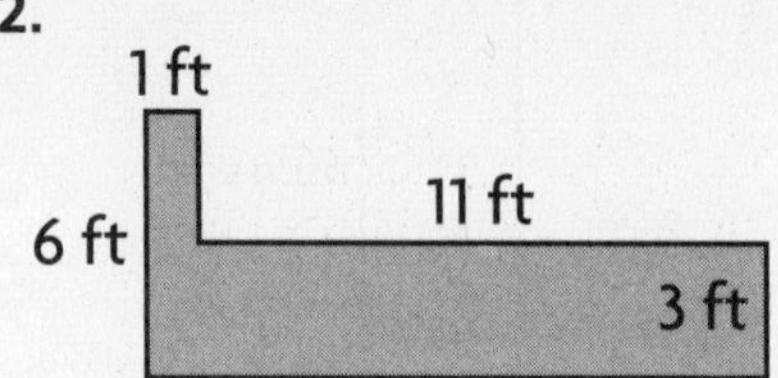

____

____

3.

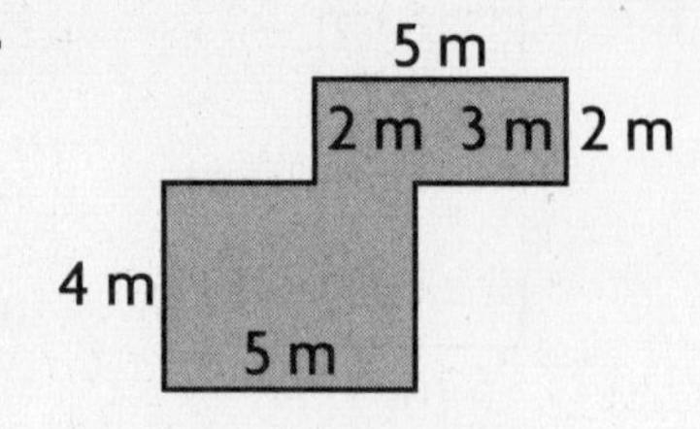

____

____

4.

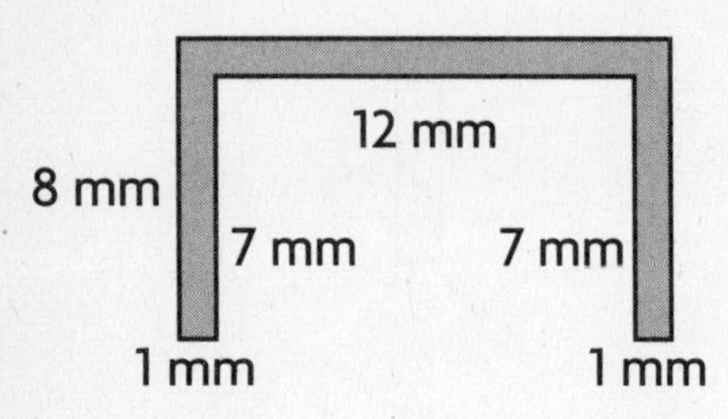

____

____

5.

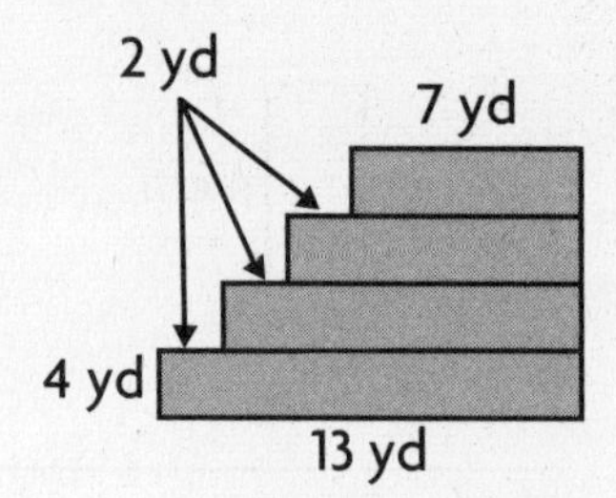

____

____

6.

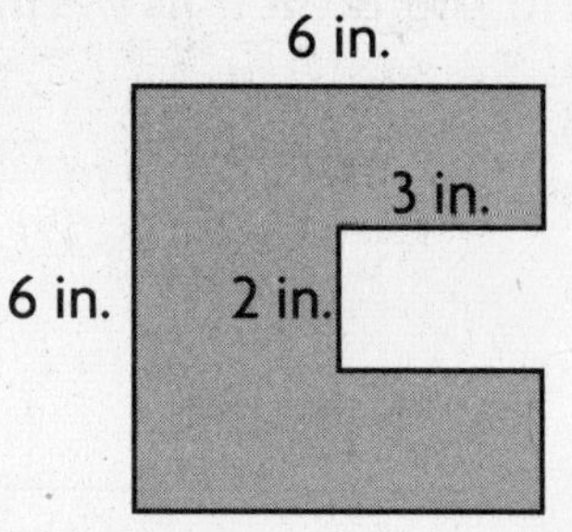

____

____

Problem Solving and TAKS Prep

For 7–8, use Exercises 5 and 6 above.

7. Look at Exercise 5. What would be the perimeter if a fifth step that is 5 yd long were added to the top?

____

8. Look at Exercise 6. What would be the area if all of the measures were doubled?

____

9. What is the total perimeter of the figure below?

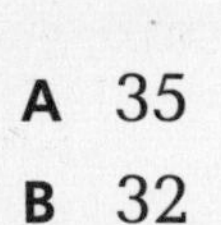

A 35

B 32

C 28

D 44

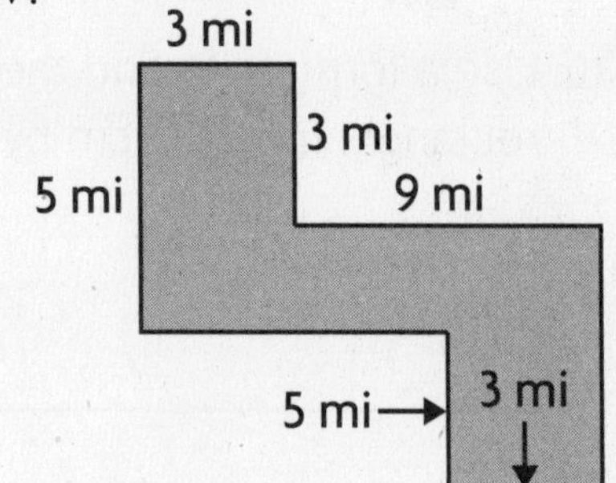

10. What is the total area of the figure in Exercise 9?

F 30

G 34

H 48

J 42

Name_______________________________ Lesson 23.8

Relate Perimeter and Area

Find the area and perimeter of each figure. Then draw another figure that has the same perimeter but a different area.

1. 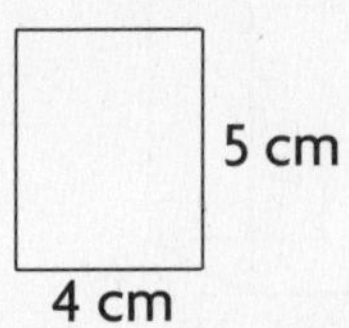

2. 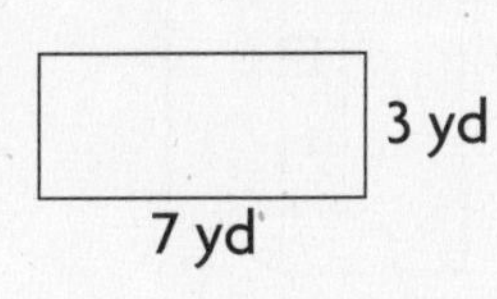

3. 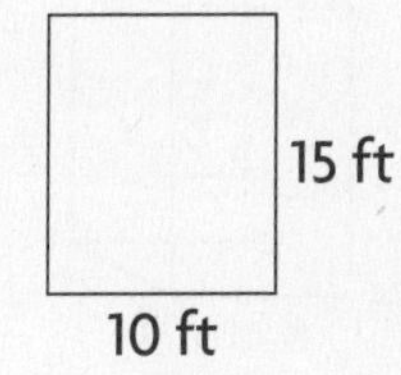

_______________ _______________ _______________

_______________ _______________ _______________

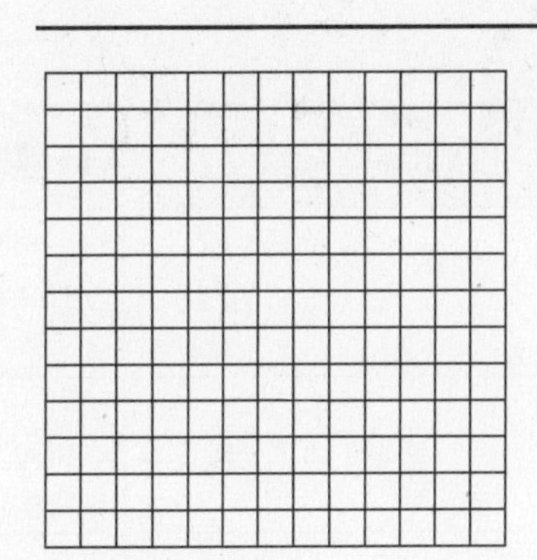

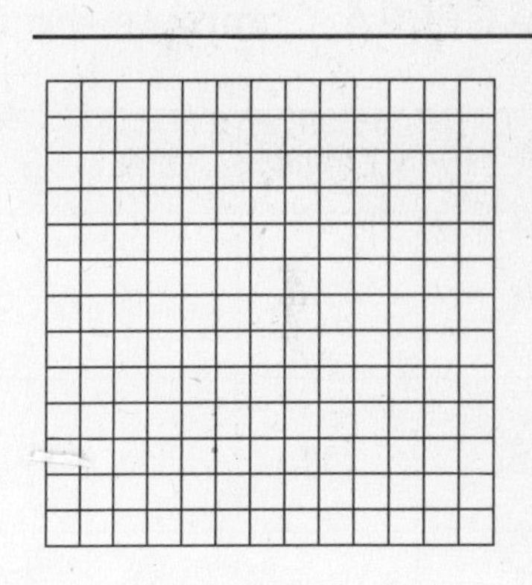

Problem Solving and TAKS Prep

For 4–5, use figures a–c.

4. Which figures have the same area but different perimeters?

____

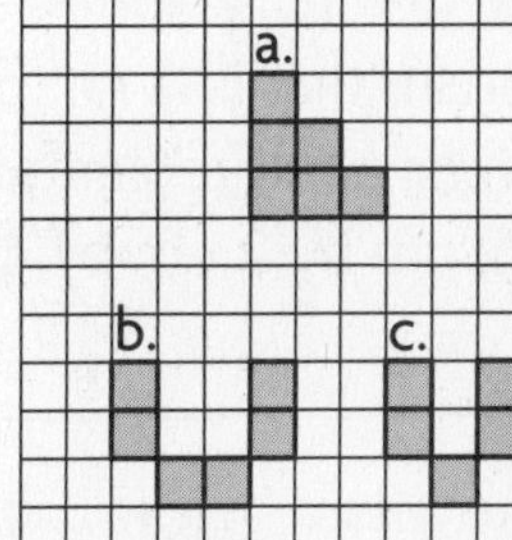

5. Which figures have the same perimeter but different areas?

____

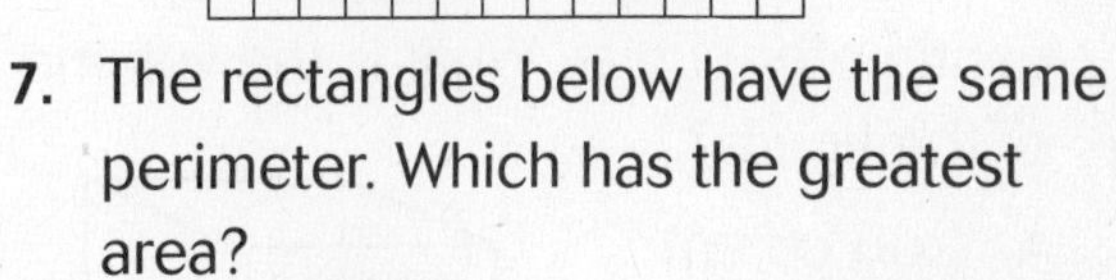

6. The rectangles below have the same area. Which has the greatest perimeter?

A 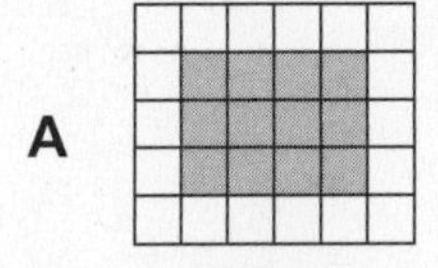

C 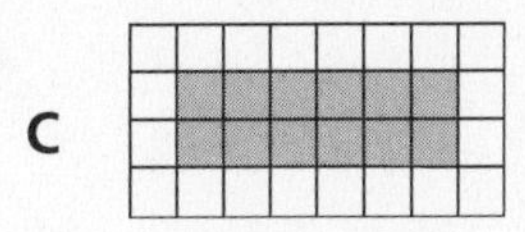

B 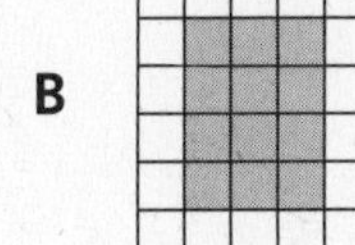

D 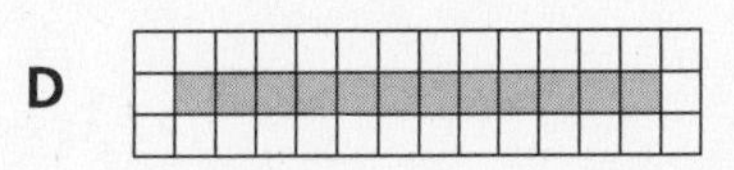

7. The rectangles below have the same perimeter. Which has the greatest area?

F 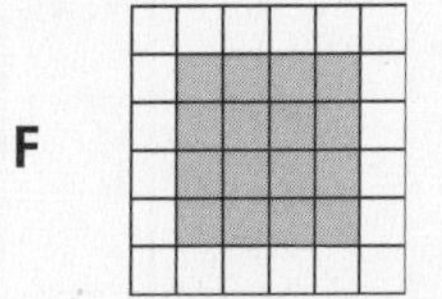

H 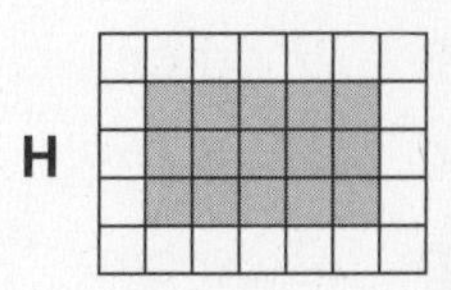

G 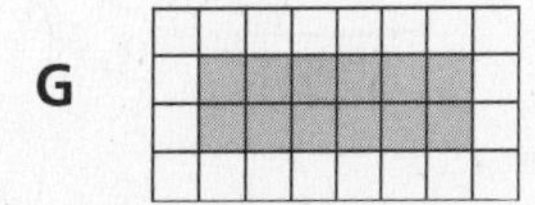

J 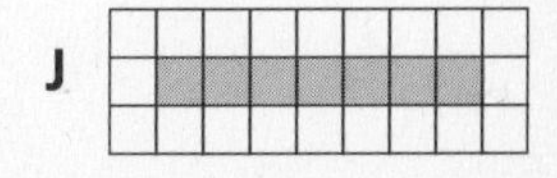

Name____________________

Estimate and Measure Volume of Prisms

Estimate. Then count to find the volume.

1.

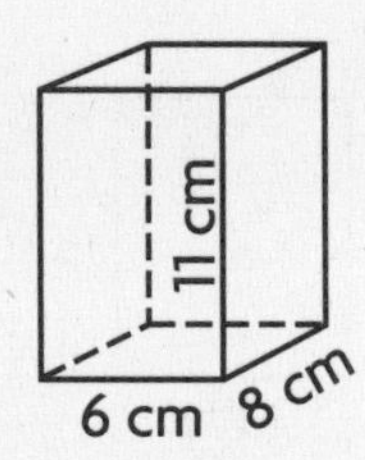

____

2.

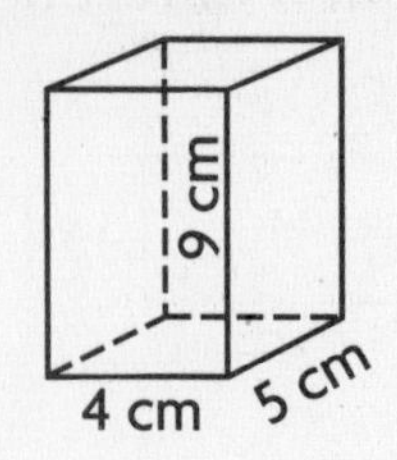

____

3.

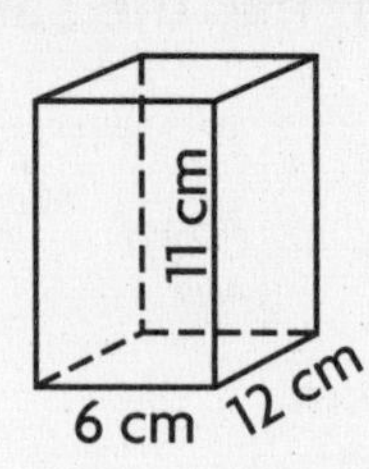

____

ALGEBRA Complete the table.

	Length	Width	Height	Volume		Length	Width	Height	Volume
4.	4 cm	3 cm	7 cm		**7.**	15 m	10 m		900 cu m
5.	6 in.	3 in.	9 in.		**8.**	11 yd		4 yd	220 cu yd
6.	12 ft	7 ft	15 ft		**9.**		12 mm	15 mm	2,700 cu mm

Problem Solving and TAKS Prep

10. Which has the greater volume, a blue rectangular prism that is 4 cm by 1 cm by 3 cm or a red rectangular prism that is 2 cm by 2 cm by 4 cm?

____

11. The volume of a rectangular prism is 200 cubic cm. If the length and width are each 5 cm, what is the height?

____

12. Jamal built the prism below using centimeter cubes. What is the volume of the figure?

A 288 cu cm

B 72 cu cm

C 48 cu cm

D 24 cu cm

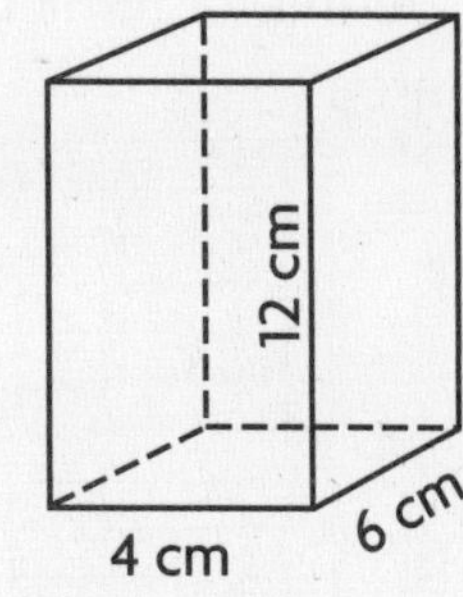

13. The volume of a rectangular prism is 60 cubic inches. The height is 4 inches. What could the length and width of the rectangular prism be?

____

Name ______________________________

Problem Solving Workshop Skill: Too Much/Too Little Information

Problem Solving Skill Practice

Decide whether the problem has too much or too little information. Then solve, if possible.

1. Jana is filling a package to send to her grandmother. The package is 8 in. by 12 in. and filled with 20 items. If the volume of the package is 1,440 cu in., what is the height of the package?

____

2. Georgia is making a rectangular prism for her 60-pound dog to sleep in. One cubic yard is 46,656 inches. The volume of the rectangular prism is 2 cu yd. If the length and width are each 1 yd, can a dog that is 30 inches walk into the prism standing upright?

____

3. Garth and three cousins are building a derby racer using 5 wheels, a disc, and a wooden box that is 2 ft wide by 1 yd high by 4 feet long. What is the volume of the box?

____

4. Monica's jewelry box is 6 in. by 5 in. It has 2 drawers. The box is painted pink and has 7 navy blue ribbons. What is the volume of the jewelry box?

____

Mixed Applications

5. **Pose a Problem** Use the information in Exercise 4. Rewrite the problem by changing the numbers so that it can be solved.

____

____

6. An amusement park has 12 bumper cars. Each car can hold 3 people. A bumper car ride lasts 10 minutes. If all the cars are filled all the time, how many people can ride in an hour?

____

7. Cyndi needs 55 glass gems to decorate a cube. Gems come 7 in a package for $3.79. How much will Cyndi need to spend to decorate her cube?

____

8. **Open Ended** Jill is making stackable cubes. How will the dimensions of each cube relate to one another?

____

____

Name____

Lesson 24.3

Compare Volume of Prisms

Compare the volume of the figures. Write <, >, or = for each ●.

1.

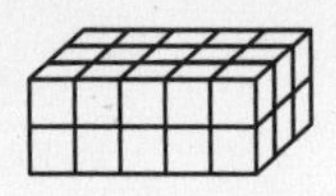  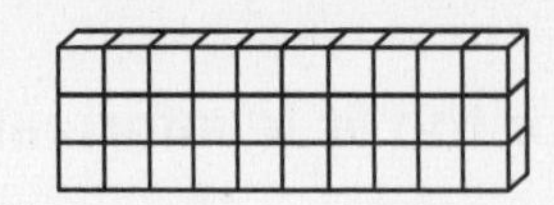

2.

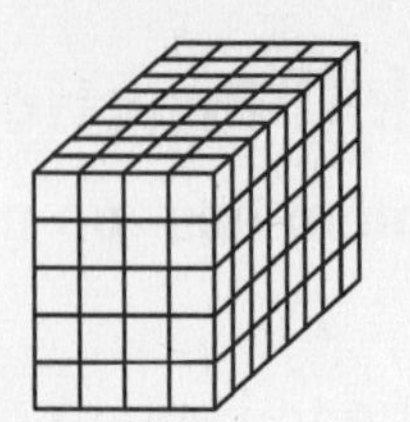  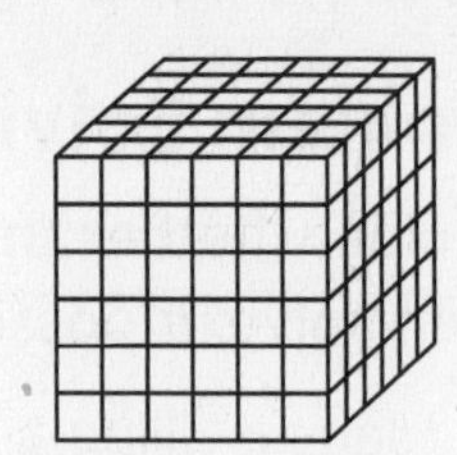

Build or draw each prism. Then compare the volumes.

3. Prism A:
length: 10 cm
width: 3 cm
height: 7 cm

Prism B
length: 10 cm
width: 5 cm
height: 4 cm

Prism A ● Prism B

4. Prism A:
length: 5 cm
width: 8 cm
height: 6 cm

Prism B
length: 6 cm
width: 6 cm
height: 5 cm

Prism A ● Prism B

5. Prism A:
length: 2 cm
width: 2 cm
height: 5 cm

Prism B
length: 5 cm
width: 4 cm
height: 1 cm

Prism A  Prism B

6. Prism A:
length: 4 cm
width: 3 cm
height: 6 cm

Prism B
length: 2 cm
width: 3 cm
height: 12 cm

Prism A ● Prism B

Problem Solving and TAKS Prep

7. Which has the greater volume, a prism 4 units long, 5 units wide, and 6 units high or a cube with a length, width, and height of 6 units?

____

8. Two prisms have the same volume. One is 4 mm by 6 mm by 10 mm. The other is 8 mm long and 15 mm high. How wide is the second prism?

____

9. Amie has a cube shaped toy chest. The volume is 125 cubic inches. What could be the length, width, and height of the toy chest?

____

10. Carey has box of miniature figures that has a volume of 343 cubic inches. What could be the length, width, and height of the box?

____

Name____

Lesson 24.4

Relate Volume and Capacity

Tell whether you would measure the *capacity* or the *volume*.

1. the amount of water in an ice cube tray

____

2. the amount of space taken up by a thermos

____

3. the amount of air in a tire

____

4. the amount of tea in a cup

____

5. the amount of space in a truck cab

____

6. the amount of space in a room

____

7. the amount of oil in a pipeline

____

8. the amount of blood in human veins

____

Problem Solving and Test Prep

USE DATA For 9–10, use the table.

Object	Capacity	Volume
Topsoil	26 quarts	1 cubic foot
Soda	2 liters	61 cubic inches
Milk carton	1 quart	About 58 cubic inches

9. How many quarts of topsoil can be held in a 5-cubic foot container?

____

10. About how many cubic inches will it take to store 10 quarts of milk?

____

11. Cass wants to find out how much gas will fit in the gas tank of the family car. Which measurement should she find?

A capacity
B length
C mass
D volume

12. Lex wants to find out how much hot cocoa his thermos holds. Which measurement should he find?

F capacity
G length
H mass
J volume

SPIRAL REVIEW

Spiral Review

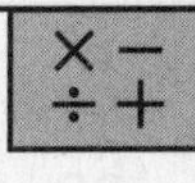

For 1–4, compare the numbers. Write <, >, or = for each ◯.

1. 56 ◯ 79
2. 324 ◯ 423
3. 912 ◯ 912
4. 203 ◯ 193

For 5–8, write the time. Write one way you can read the time.

5. 

6. 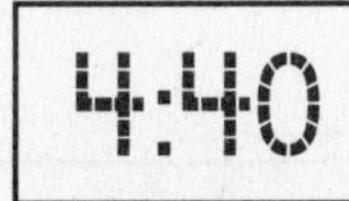

7. 

8. 

For 9–11, use the data in the graph.

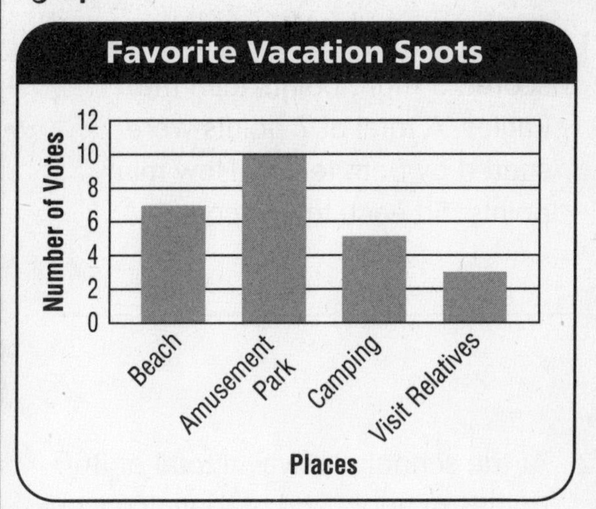

9. How many people voted for the beach? ______
10. How many people voted for the amusement park? ______
11. Which place had the least amount of votes?

____

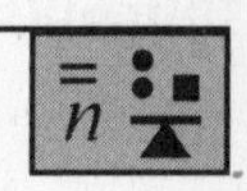

For 12–13, write the fact family for each set of numbers.

12. 2, 3, 5

13. 2, 7, 9

Name ______________________________ **Week 2**

Spiral Review

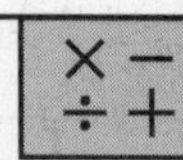

For 1–2, use predict and test to solve the problems.

1. During a soccer game, the Dragons scored 3 more points than the Eagles. A total of 7 points were scored by both teams. How many points did each team score?

____

2. At the school carnival, a total of 300 bottles of apple and grape juice were sold. There were 80 more bottles of apple juice sold than grape juice. How many bottles of each type of juice were sold?

____

For 6–8, use the data in the table.

Mrs. Yin's class voted for their favorite color.

Color	Votes
Black	2
Blue	8
Green	5
Red	7
Yellow	4

6. Which color had the most votes?

____

7. How many votes were there in all?

____

8. How many more students liked red than yellow?

____

For 3–5, choose the better unit of measure.

3. The length of a school bus: 9 feet or 9 yards?

____

4. The distance between New York City and Los Angles: 2,500 yards or 2,500 miles?

____

5. The amount of coffee in a mug: 2 cups or 2 quarts?

____

For 9–11, name each figure as a *line*, a *ray*, or a *line segment*.

9. ⟷

____

10. •——•

____

11. •——→

____

Name ______________________ **Week 3**

Spiral Review

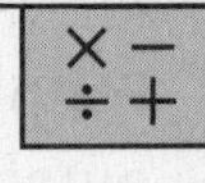

For questions 1–4, compare using <, >, or = in each ◯.

1. 5,327 ◯ 5,341

2. 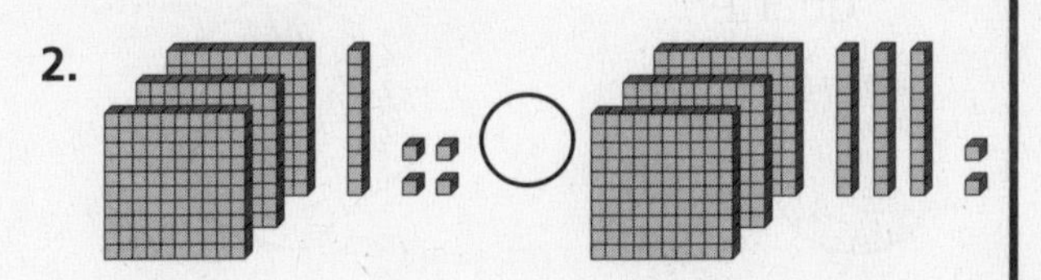

3. 3,300 3,340 3,380 3,420 3,460

   3,340 ◯ 3,460

4. 4,039 ◯ 4,039

For 5–8, choose the unit you would use to measure each. Write *cup, pint, quart,* or *gallon*.

5. ____________ 6. ____________

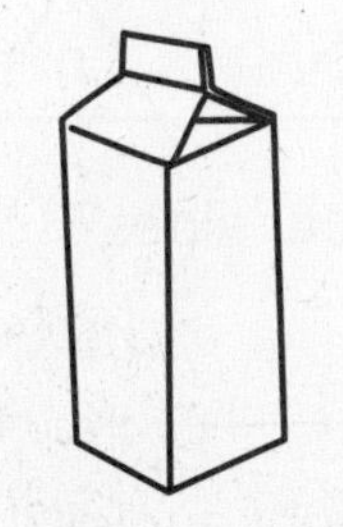

7. ____________ 8. ____________

For 9–11, use the table below to tell whether each event is *likely, unlikely,* or *impossible*.

Warren's Marbles	
Blue	●●
Green	●●●●●●●●
Red	●●●●●●●●●●●●●●

9. Warren will pull out a red marble.

   ____

10. Warren will pull out a yellow marble.

    ____

11. Warren will pull out a blue marble.

    ____

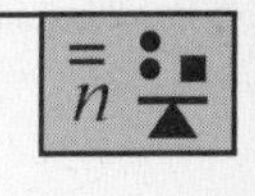

For 12–15, write a related fact. Use it to complete the number sentence.

12. 7 + ☐ = 9

    ____

13. ☐ − 3 = 9

    ____

14. 4 + ☐ = 6

    ____

15. 8 − ☐ = 5

    ____

Name ______________________ **Week 4**

Spiral Review

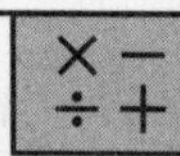

For 1–2, tell whether to estimate or find an exact answer. Then solve the problem.

1. Ketan needs to buy a notebook, a pencil, and a pen. A notebook costs $4.75, a pencil costs $1.29, and a pen costs $1.69. About how much money does Ketan need?

____

____

2. A school bus has 32 rows of seats. Two students can sit in each seat. How many students can the school bus carry?

____

____

For 3–5, choose the best estimate.

3. Diameter of a penny 2 cm or 2 dm

____

4. Length of a school book 300 cm or 3 dm

____

5. Length of a hairbrush 15 cm or 15 dm

____

For 6–9, list the possible outcomes for each.

6. Elena flips a quarter

____

____

7. Gene rolls a die

____

____

8. Lynne spins a pointer.

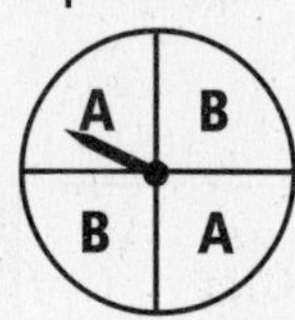

____

____

9. Haley picks a marble

____

____

For 10–12, use the figures below.

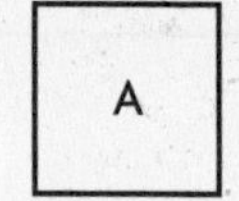

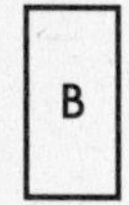

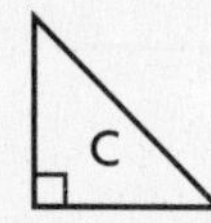

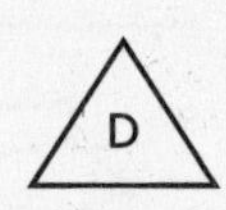

10. Which figures have four sides?

____

11. Which figures have right angles?

____

12. Which figure has three acute angles?

____

Spiral Review

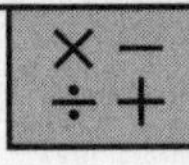

For 1–4, estimate. Then the find the sum or difference.

1. 2,345 + 1,179

____

____

2. 4,845 − 2,954

____

____

3. 9,678 − 928

____

____

4. 6,429 + 3,218

____

____

For 7–8, use the graph below.

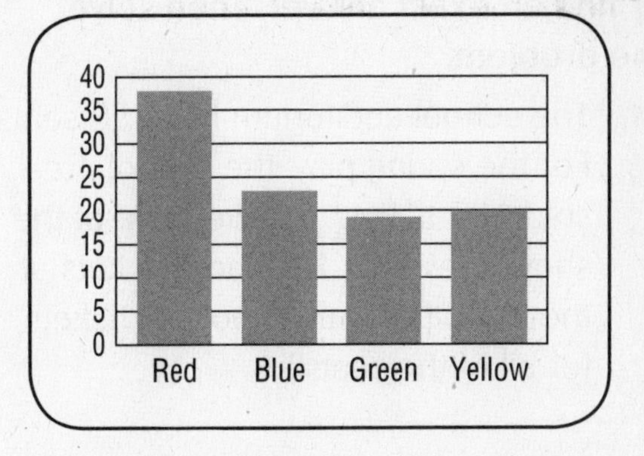

7. Kimeiko is going to spin a spinner. On which color will the spinner likely stop?

____

8. Is the number of times the spinner landed on red greater or less than the number of times for green?

____

For 5–6, use cubes to make each solid. Then write the volume in cubic units.

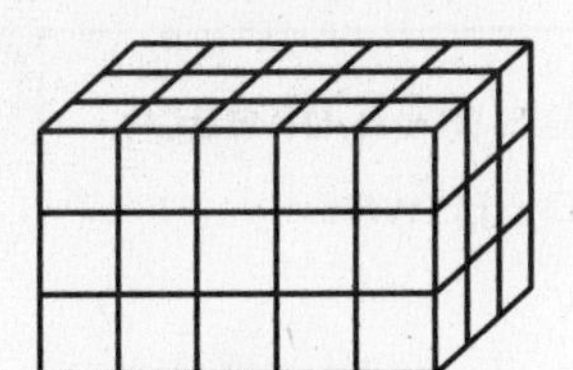

5. ______________________

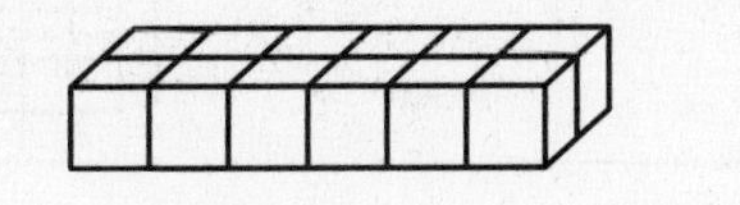

6. ______________________

For 9–12, find a possible pattern. Then predict the next two numbers or shapes in each pattern.

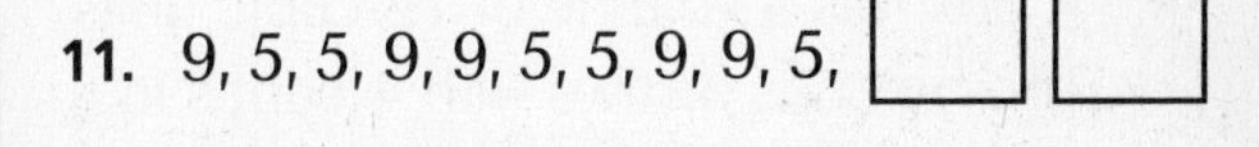

9. ●□□●□□●□□ ☐ ☐

10. 2, 2, 2, 5, 2, 2, 2, 5, 2, ☐ ☐

11. 9, 5, 5, 9, 9, 5, 5, 9, 9, 5, ☐ ☐

12. □□▲▲□□▲▲□□ ☐ ☐

Spiral Review

For 1–2, tell whether to estimate or find an exact answer. Then solve the problem.

1. The school auditorium has 300 seats. For the spring play, the school pre-sold 187 tickets. On the night of the show, they sold 109 more tickets at the door. Did the school sell tickets for all of the seats?

   ____

2. Caesar's track coach wants him to run about 15 miles a week. If Caesar runs 4 miles on Monday, 7 miles on Wednesday, and 3 miles on Friday, will he have run enough?

   ____

For 3–4, use a model to solve.

3. Ian filled a box with blocks. There were 4 layers. Each layer had 2 rows of 4 blocks. What was the volume of the box?

   ____

4. A box has a volume of 12 cubic units. The box has 3 rows with 4 cubes in each row. How many layers does the box have?

   ____

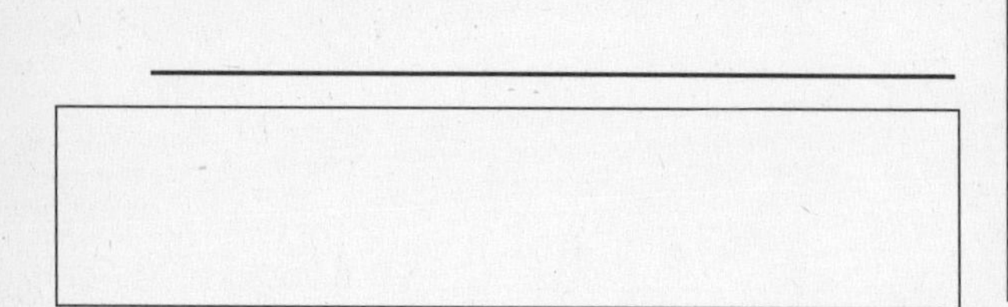

For 5–7, use the Favorite Fruit pictograph.

5. Use the data in the pictograph to make a bar graph.

6. Which fruit had the most votes?

   ____

7. How many more votes did bananas get than oranges?

   ____

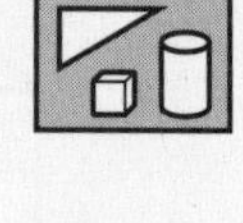

For 8–10, find the area of each figure. Write the answer in square units.

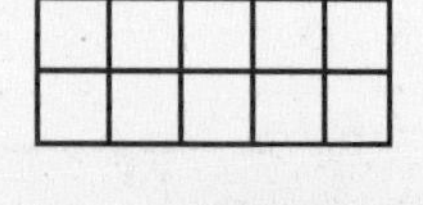

8. ____________________

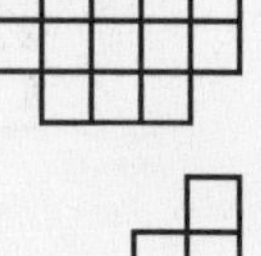

9. ____________________

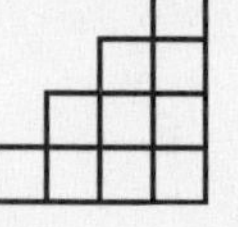

10. ____________________

Name ______________________________ **Week 7**

Spiral Review

For 1–4, find the product. 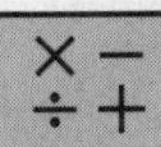

1. $4 \times 3 =$ ______

2. $8 \times 4 =$ ______

3. 5×4 = ______

4. 4×4 = ______

For 5–6, write the time. Write one way you can read the time. 

5.

____

____

6.

____

____

____

For 7–9, use the survey below. 

Students' Hair Color in Mr. Provost's Class	
Black	6
Blonde	5
Brown	8
Red	2

7. What color of hair was most common? ______

8. How many students were observed?

____

9. What is the title of this survey?

____

____

____

For 10–13, find the value of the variable. Then write a related sentence. 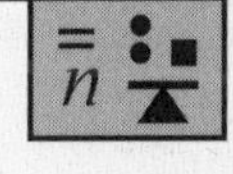

10. $36 \div t = 6$

____

11. $a \times 3 = 21$

____

12. $y \div 3 = 6$

____

13. $9 \times m = 63$

____

Name ______________________________ **Week 8**

Spiral Review

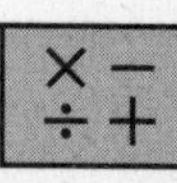

For 1–4, find the quotient. Write a related multiplication sentence.

1. $120 \div 10 =$ ______

____

2. $99 \div 11 =$ ______

____

3. $36 \div 3 =$ ______

____

4. $72 \div 2 =$ ______

____

For 5–8, write each temperature in °F.

5. 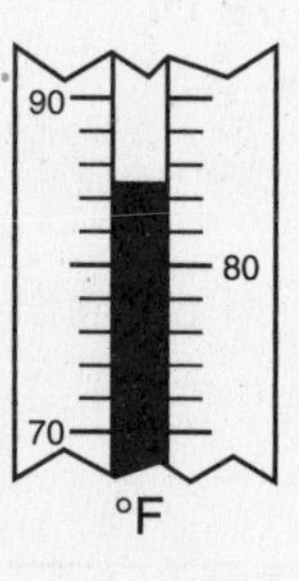

______

6. 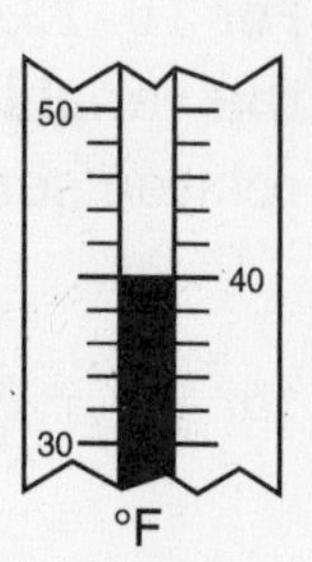

______

7. 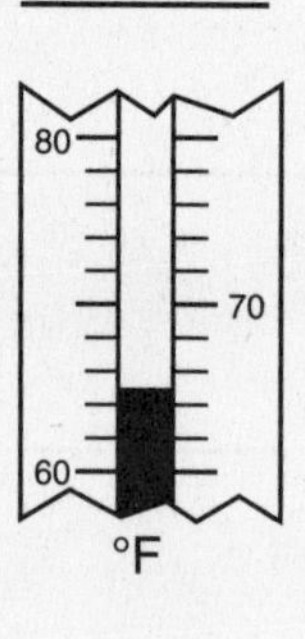

______

8. 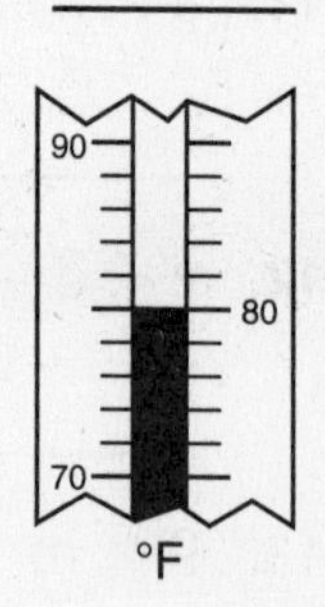

______

For 9–11, use the pictograph below.

Touchdowns Scored in a Season	
Ryan	🏈🏈🏈🏈
Eric	🏈🏈
Roy	🏈🏈
Each 🏈 = 2 touchdowns.	

9. How many touchdowns did Ryan score?

____

10. Which players scored the same number of touchdowns?

____

11. How many touchdowns were scored in all?

____

For 12–14, trace and cut out each pair of figures. Tell if the figures are congruent. Write *yes* or *no*.

12. ______ 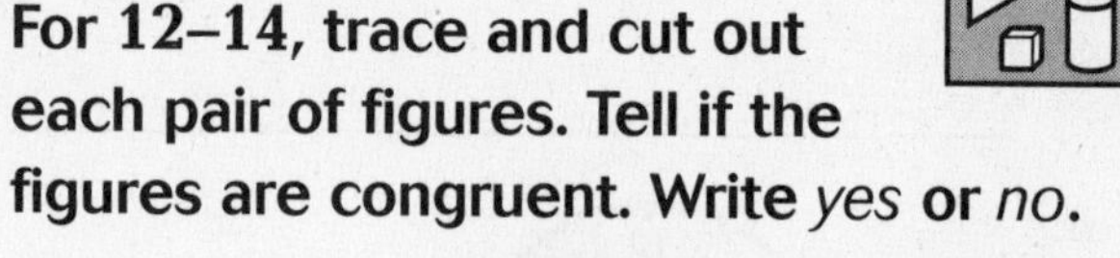

13. ______ 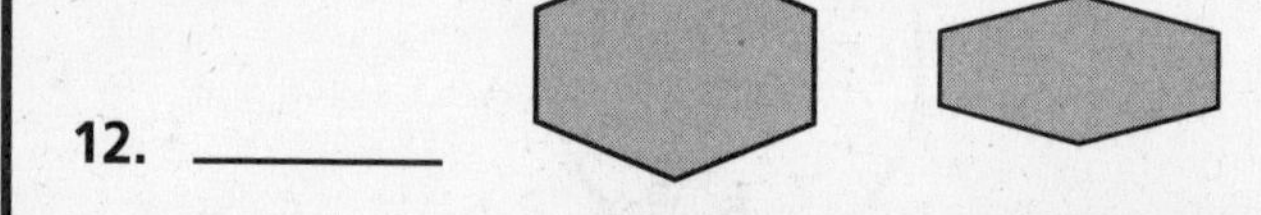

14. ______ 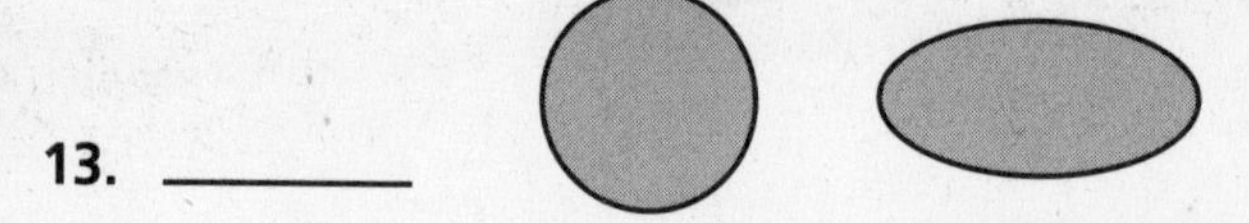

Spiral Review

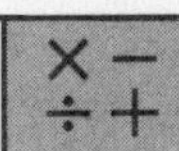

For 1–3, find the product.

1. $5 \times 17 = \square$

2. $5 \times 9 = \square$

3. $10 \times 13 = \square$

For 4–7, use a ruler. Draw a line for each length.

4. 1 inch

5. $2\frac{1}{4}$ inches

6. $1\ 1\frac{1}{2}$ inches

7. $1\frac{1}{8}$ inches

For 8–10, use the table below.

Shirt Size	Boys	Girls
Small	8	7
Medium	6	9
Large	4	2

8. How many boys wear medium?

____

9. How many students were surveyed in all?

____

10. How many more girls wear medium than boys?

____

For 11–13, use the rule and equation to make an input/output table.

11. multiply by 4; $t \times 4 = h$

Input					
Output					

12. divide by 6; $u \div 6 = z$

Input					
Output					

13. multiply by 10; $b \times 10 = c$

Input						
Output						

Spiral Review

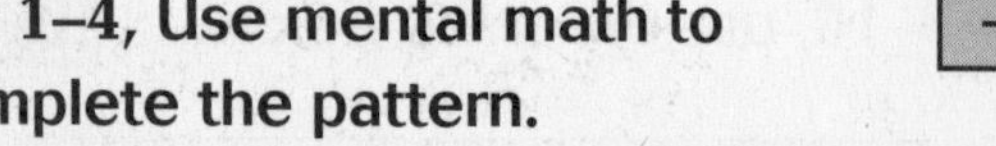

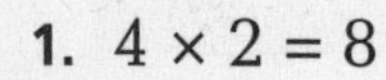

For 1–4, Use mental math to complete the pattern.

1. $4 \times 2 = 8$

$4 \times 20 = \square$

$4 \times 200 = \square$

$4 \times 2{,}000 = \square$

2. $6 \times 8 = 48$

$6 \times 80 = \square$

$6 \times 800 = \square$

$6 \times 8{,}000 = \square$

3. $8 \times 3 = 24$

$8 \times 30 = \square$

$8 \times 300 = \square$

$8 \times 3{,}000 = \square$

4. $5 \times 7 = 35$

$5 \times 70 = \square$

$5 \times 700 = \square$

$5 \times 7{,}000 = \square$

For 5–8, compare. Write <, >, or = for each ◯.

5. 19 ounces ◯ 1 pound

6. 32 ounces ◯ 2 pounds

7. 23 ounces ◯ 3 pounds

8. 8 ounces ◯ 1 pound

For 9–11, use the chart below to answer the questions.

Number of Books Read by Mr. Hern's Class

0 1 2 3 4 5 6 7 8 9

Jan. Feb. Mar. Apr.

9. How many books were read in all?

____

10. In which months were the same number of books read?

____

11. How many more books were read in January than in April?

____

For 12–14, tell if the line is a line of symmetry. Write *yes* or *no*.

12. ________

13. ________

14. ________

Name ______________________________ **Week 11**

Spiral Review

For 1–2, solve the problem.

1. In 1996, people in the United States ate 100 acres of pizza every day. Would it be reasonable to say that people eat 1,000 acres of pizza a week? Explain.

____

____

2. The Paine family is having a pizza party. They need 9 pizzas. If each pizza cost \$17, is \$150 a reasonable estimate for the total? Explain.

____

For 3–4, use the thermometers. Find the difference in temperatures.

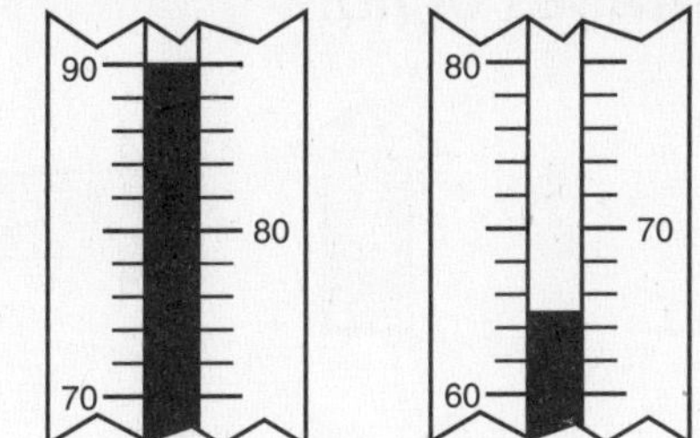

3. __________

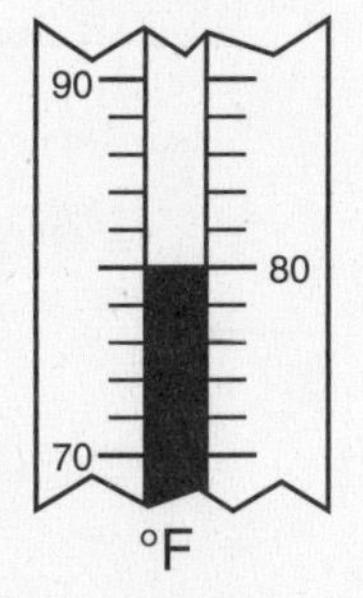

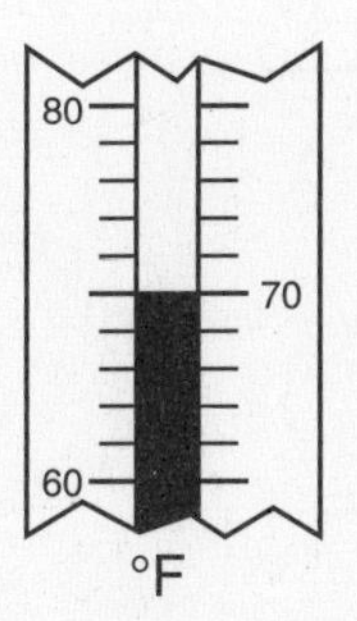

4. __________

For 5–7, use the spinner below.

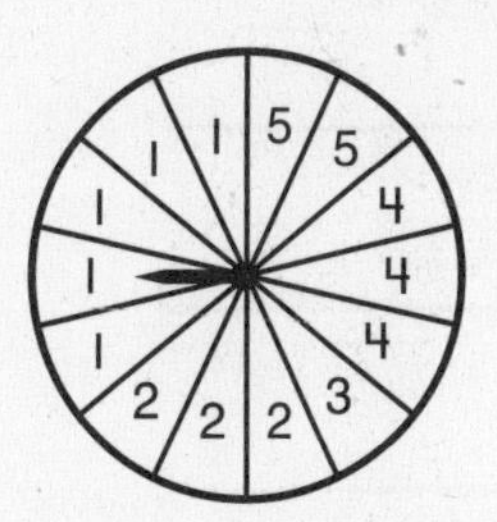

5. Which outcome is most likely?

____

6. Which outcome is least likely?

____

7. Which outcomes are equally likely?

____

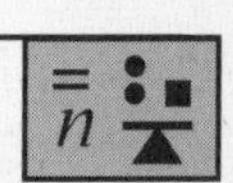

For 8–11, find the value of the variable. Then write a related sentence.

8. $36 \div t = 6$

____

9. $a \times 3 = 21$

____

10. $y \div 3 = 6$

____

11. $9 \times m = 63$

____

Spiral Review

For 1–5, estimate the product.

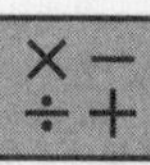

1. $78 \times 34 =$ __________
2. $91 \times 46 =$ __________
3. $22 \times 33 =$ __________
4. $61 \times 359 =$ __________
5. $20 \times 119 =$ __________

For 9–10, use the graph below.

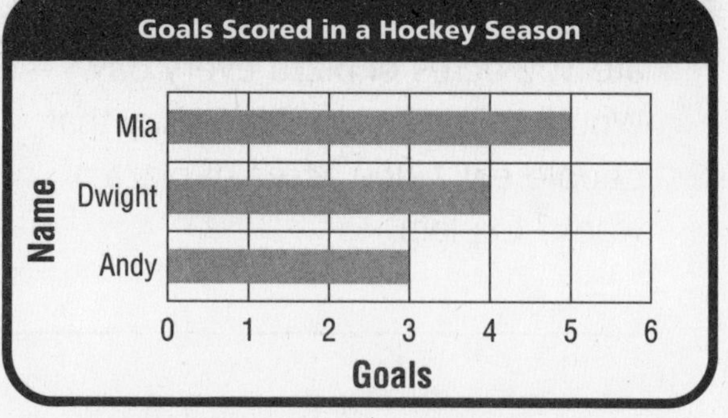

9. How many total hockey goals were scored for the entire season?

____

10. How many more goals did Mia score than Andy?

____

For 6–8, choose the unit you would use to measure each. Write *centimeter*, *meter*, or *kilometer*.

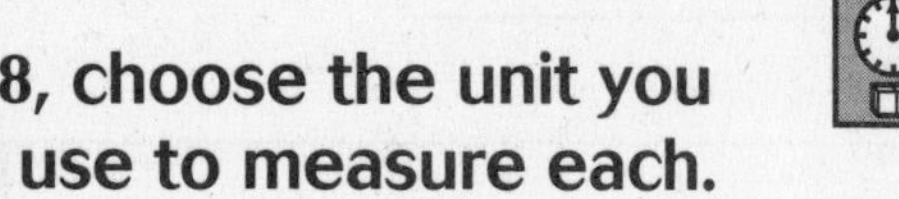

6. height of a school desk

____

7. length of a pencil

____

8. four city blocks

____

For 11–13, Tell if a translation was used to move the figure. Write *yes* or *no*.

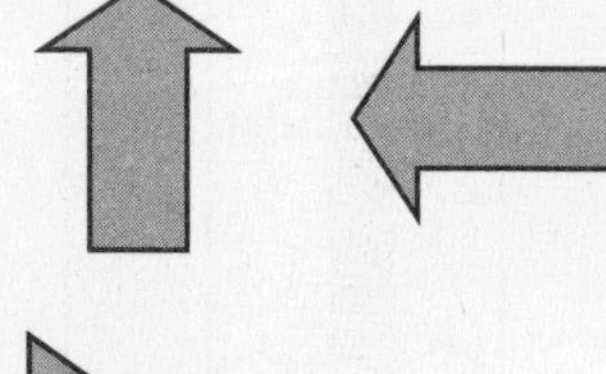

11. __________

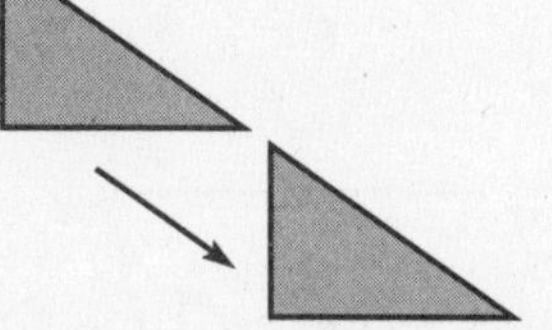

12. __________

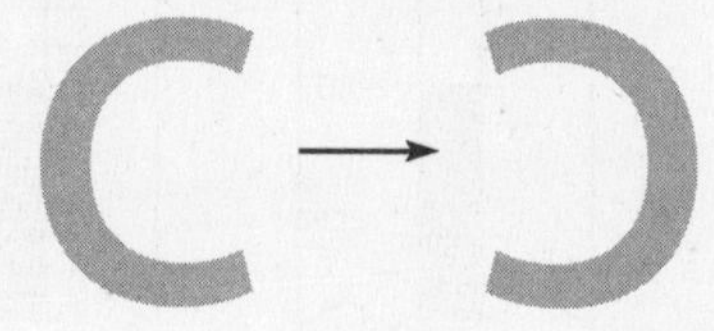

13. __________

Spiral Review

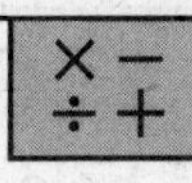

For 1–5, round each number to the place value of the underlined digit.

1. $1,\underline{7}54$ ____________
2. $\underline{4}5,981$ ____________
3. $7\underline{1}3,402$ ____________
4. $\underline{3},922,703$ ____________
5. $9,\underline{7}79,911$ ____________

For 6–9, compare.

6. Which is wider, a sheet of notebook paper or a doorway?

____

7. Which is heavier, a school book or a paper clip?

____

8. Which holds more, a bathtub or a swimming pool?

____

9. Which is longer, a yardstick or a pencil?

____

For 10–12, use the data in the pictograph.

Fish Caught

Aidan	🐟🐟🐟🐟🐟🐟
Steve	🐟🐟
Tracy	🐟🐟🐟🐟

Key: Each 🐟 = 2 fish

10. Who caught the most fish?

____

11. Who caught the least fish?

____

12. How many fish were caught in all?

____

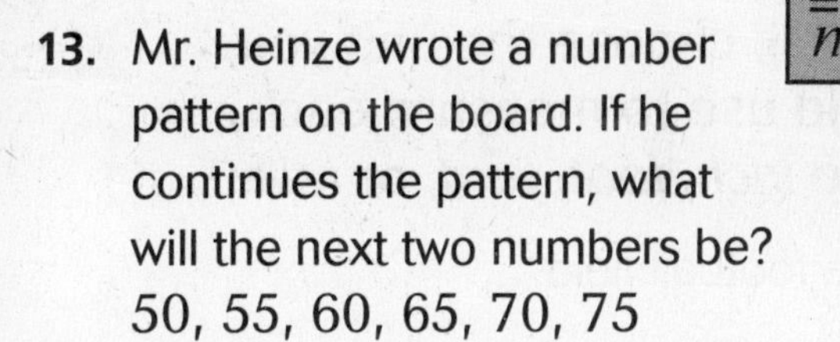

13. Mr. Heinze wrote a number pattern on the board. If he continues the pattern, what will the next two numbers be?
50, 55, 60, 65, 70, 75

____

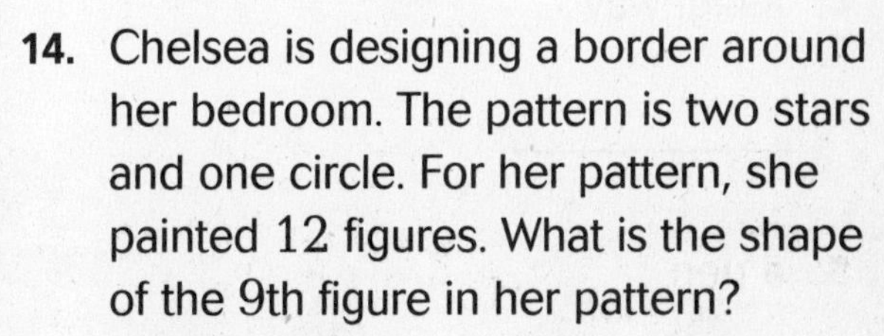

14. Chelsea is designing a border around her bedroom. The pattern is two stars and one circle. For her pattern, she painted 12 figures. What is the shape of the 9th figure in her pattern?

____

Spiral Review

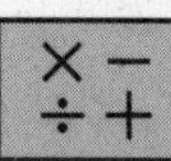

For 1–2, write a fraction in numbers and in words that names the shaded part.

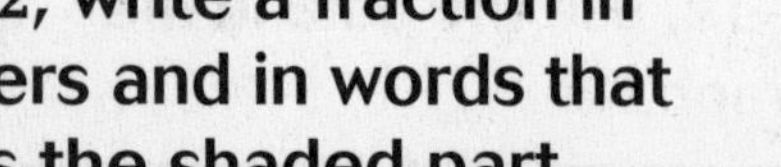

1. 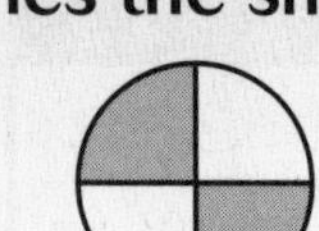

____

____

2.

____

____

For 3–6, choose the unit you would use to measure each. Write *inch*, *foot*, *yard*, or *mile*.

3. a football field

____

4. a highway

____

5. a pen

____

6. a car

____

For 7–8, use the pictograph below.

Prizes Won at the Ring Toss Game	
Stuffed Animal	★★★★
Free Game	★★★★★★★★
Goldfish	★★
Key: Each ★ = 1 prize	

7. Explain how you would display the data on the pictograph if 3 pinwheels were won.

____

____

____

8. Which prize was won the *most*? Which prize was won the *least*?

____

____

For 9–12, classify each angle as *acute*, *right*, or *obtuse*.

9. 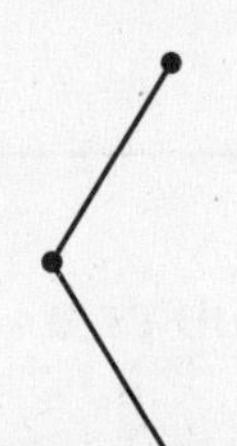

____

10. 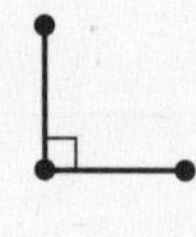

____

11. 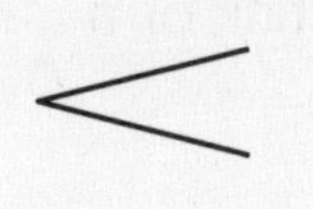

____

12. 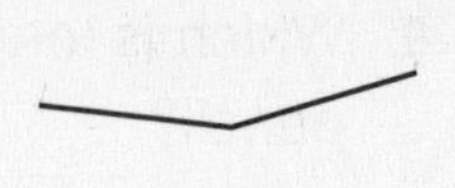

____

Spiral Review

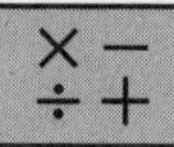

For 1–5, divide and check.

1. $498 \div 7 =$ ________
2. $186 \div 5 =$ ________
3. $304 \div 6 =$ ________
4. $101 \div 7 =$ ________
5. $\$198 \div 3 =$ ________

For 8–11, find the average.

8. 94, 84, 83, 71

____

9. 405, 323, 289

____

10. 20, 17, 12, 11, 9, 3

____

11. 78, 67, 53, 49, 31, 22

____

For 6–7, write the volume in cubic units.

6. ____________________

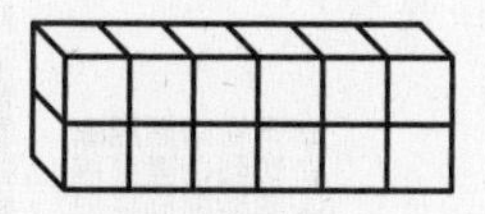

7. ____________________

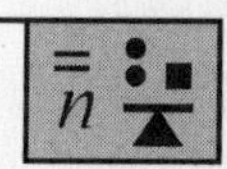

12–15, use mental math and patterns to find the product.

12. $7 \times 30 =$ ________
13. $5 \times 600 =$ ________
14. $4 \times 3{,}000 =$ ________
15. $8 \times 8{,}000 =$ ________

Spiral Review

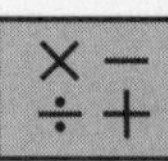

For 1–4, divide and check.

1. $189 \div 3 =$ ________

2. $564 \div 7 =$ ________

3. $898 \div 9 =$ ________

4. $732 \div 8 =$ ________

For 8–10, use the table below.

Marbles			
	Red	Blue	Green
Large	3	4	1
Medium	6	2	2
Small	1	6	0

8. How many small blue marbles are there?

____

9. How many large red marbles are there?

____

10. How many green marbles are there in all?

____

For 5–7, choose the unit you would use to measure each. Write *mL* or *L*.

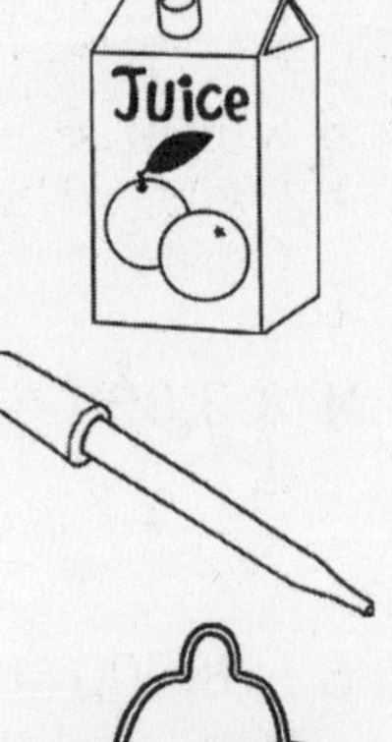

5. ________

6. ________

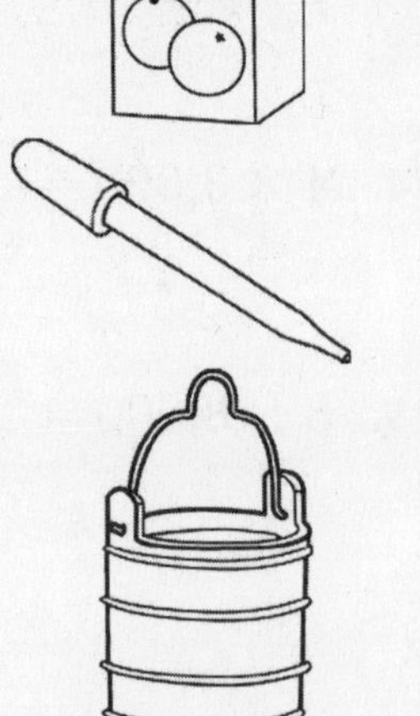

7. ________

For 11–14, write a multiplication sentence for each array.

11. ________

12. ________

13. ________

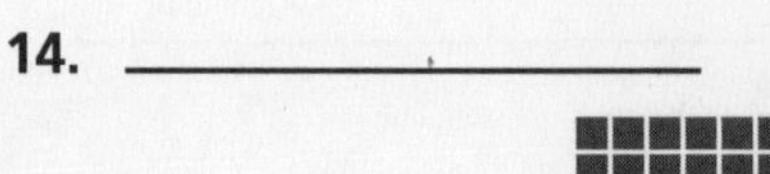

14. ________

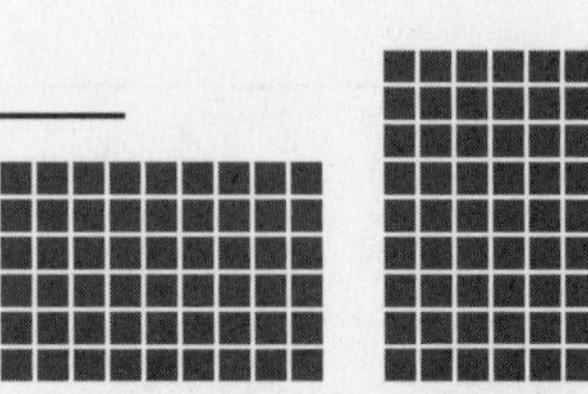

Name ______________________________

Spiral Review

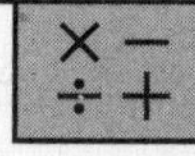

For 1–4, order the fractions from *least* to *greatest*. You may use a model.

1. $\frac{2}{3}, \frac{1}{2}, \frac{4}{5}$ ______________

2. $\frac{2}{5}, \frac{5}{8}, \frac{3}{7}$ ______________

3. $\frac{1}{3}, \frac{1}{2}, \frac{1}{4}$ ______________

4. $\frac{3}{4}, \frac{5}{6}, \frac{3}{8}$ ______________

For 8–10, use the spinner below. Tell whether each event is *likely*, *unlikely*, or *impossible*.

8. The pointer will land on 1.

____

9. The pointer will land on 3.

____

10. The pointer will land on 4.

____

For 5–7, find the perimeter of each figure.

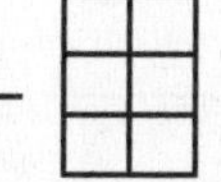

5. ______________

6. ______________

7. ______________

For 11–14, Tell how each figure was moved. Write *translation* or *rotation*.

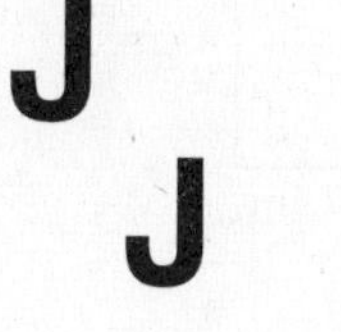

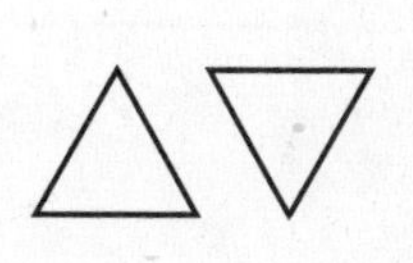

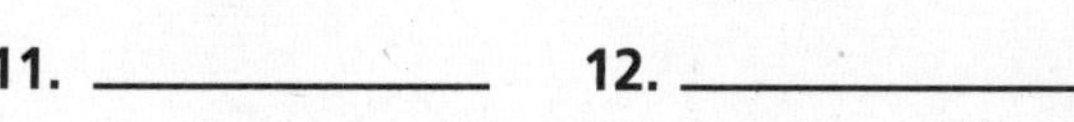

11. ______________ **12.** ______________

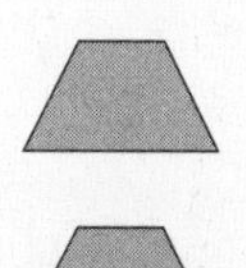

13. ______________ **14.** ______________

Spiral Review

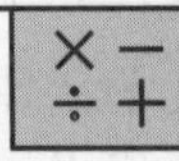

For 1–4, rename each fraction as a mixed number and each mixed number as a fraction.

1. $2\frac{3}{5}$ ______

2. $\frac{8}{3}$ ______

3. $4\frac{1}{5}$ ______

4. $\frac{15}{7}$ ______

For 8–10, list the possible outcomes for each.

8. James flips a penny

____

9. Xochi pulls a coin

____

10. Ella spins a spinner

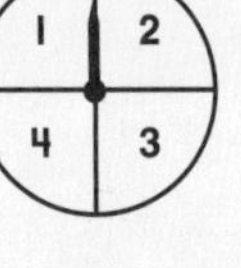
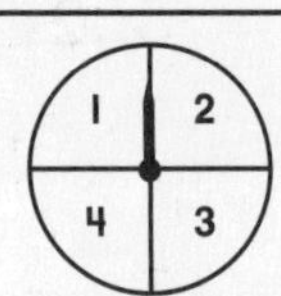

____

For 5–7, find the perimeter of the figures.

5. ______

2 cm
4 cm
4 cm
2 cm

6. ______

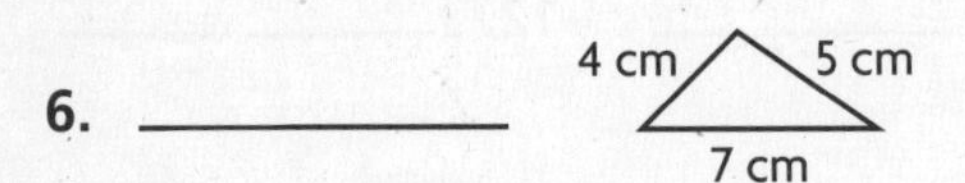

7. ______

2 cm
3 cm
5 cm
2 cm
2 cm
4 cm

For 11–14, name a solid figure that is described.

11. 6 faces

____

12. 6 edges

____

13. 5 vertices

____

14. 9 edges

____

Spiral Review

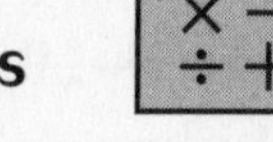

For 1–4, Use models or pictures to find the sum or difference. Record the answer.

1. $34.23 - 3.56$

2. $6.45 + 2.91$

3. $7.3 + 2.1$

4. $\$81.65 - \$49.76$

For 9–11, use the frequency table below.

Favorite Food			
	Pizza	Hamburger	Chicken Nuggets
Frequency	10	7	6

9. More students chose hamburgers than pizza. True or false?

___

10. How many students voted for chicken nuggets?

___

11. Which food got the most votes?

___

For 5–8, write the time as shown on a digital clock.

5. 25 minutes before seven

___

6. 38 minutes after nine

___

7. 10 minutes before three

___

8. 29 minutes after six

___

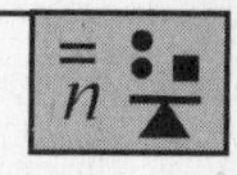

For 12–13, use logical reasoning to solve the problems.

Melinda invites Alicia, Bonnie, Carlos, and Dan to her birthday party. Alicia arrives at the party after Dan. Carlos arrives at the party after Alicia. Bonnie arrives at the party before Dan.

12. Which guest arrives first at the party?

___

13. Which guest arrives last?

___

Spiral Review

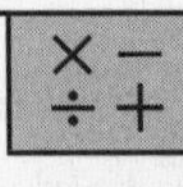

For 1–4, write each fraction as a decimal. You may use a model.

1. $\frac{2}{10}$ ________

2. $\frac{38}{100}$ ________

3. $\frac{90}{100}$ ________

4. $\frac{5}{10}$ ________

For 5–8, use the thermometer to find the temperature in °C.

5. 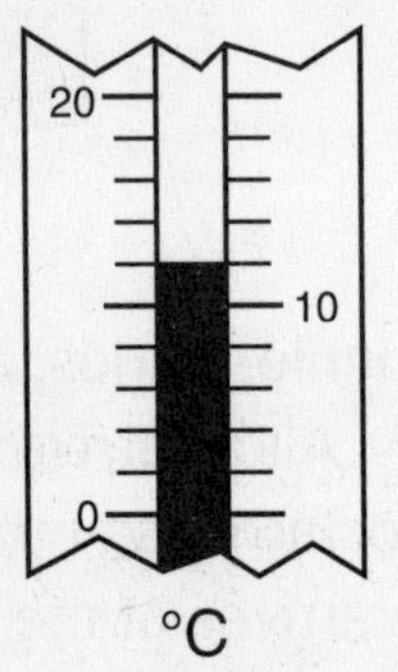

°C ________

6. 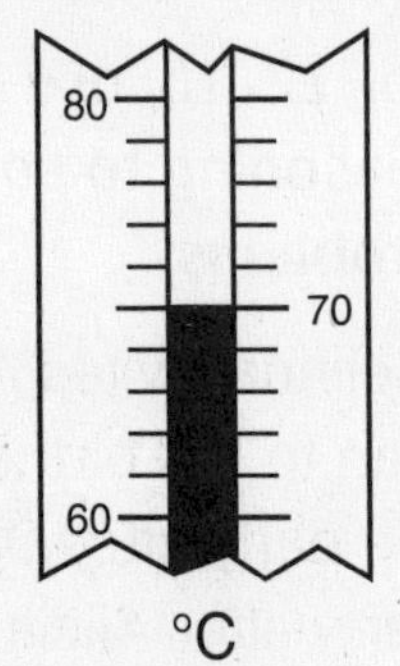

°C ________

7. 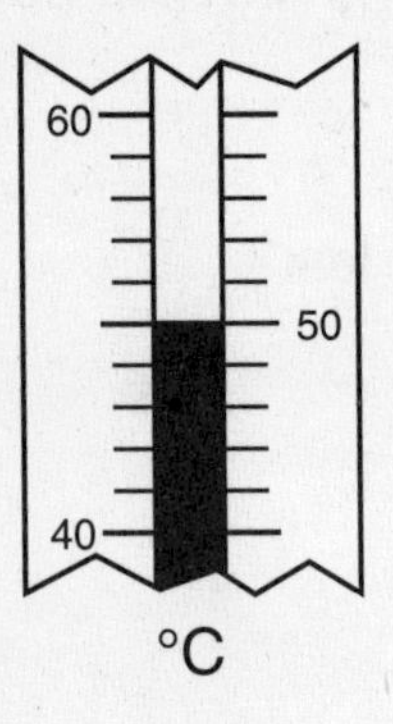

°C ________

8. 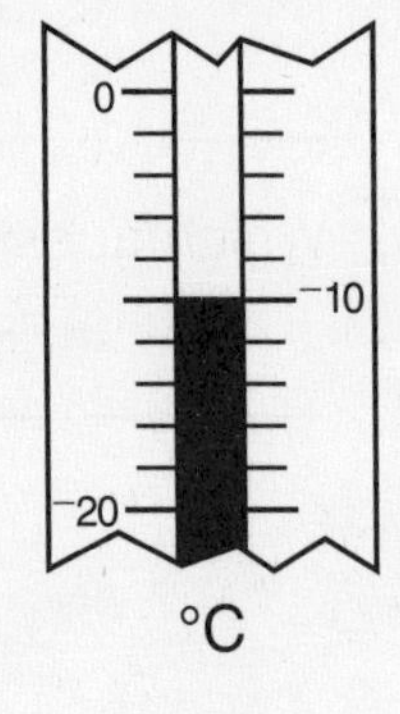

°C ________

For 9–10, use the bar graph below.

9. Which cookie was chosen by the fewest voters?

____

10. How many voters chose oatmeal as their favorite?

____

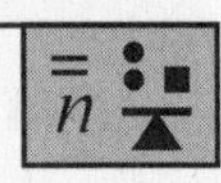

For 11–13, write a related fact. use it to complete the number sentence.

11. $7 + \square = 10$

____

12. $7 - \square = 7$

____

13. $7 - \square = 4$

____

Spiral Review

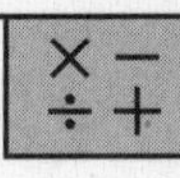

For 1–4, write two equivalent fractions for each.

1. $\frac{1}{2}$ ________

2. $\frac{8}{10}$ ________

3. $\frac{4}{6}$ ________

4. $\frac{3}{4}$ ________

For 5–8, use the thermometer to find the temperature in F.

5. ________

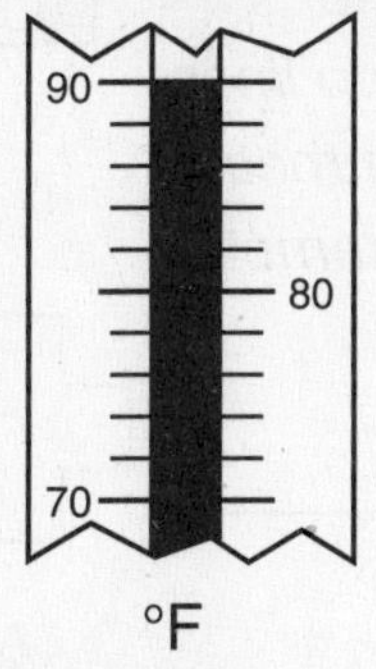

6. ________

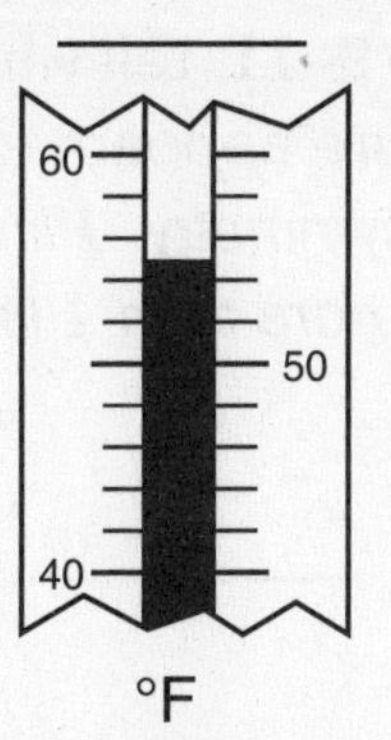

7.

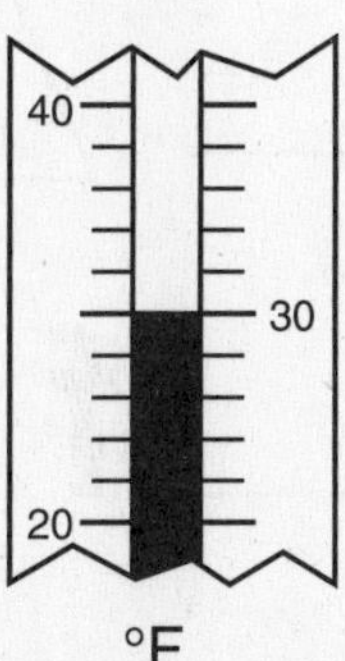

____

8.

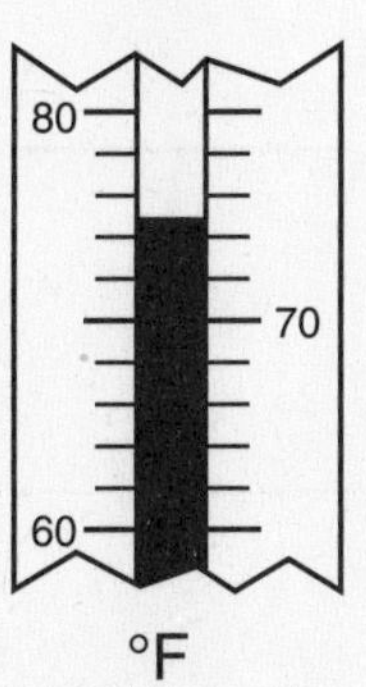

____

For 9–10, use the bar graph below.

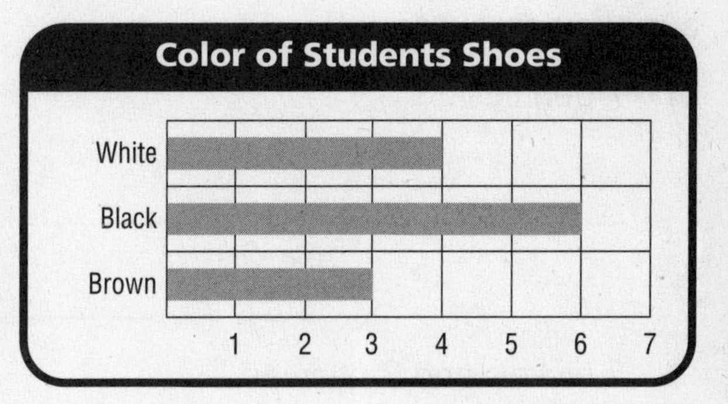

9. How many students have black shoes?

____

10. How many more students have white shoes than brown shoes?

____

For 11–13, write *intersecting, parallel,* or *perpendicular.*

11. ________

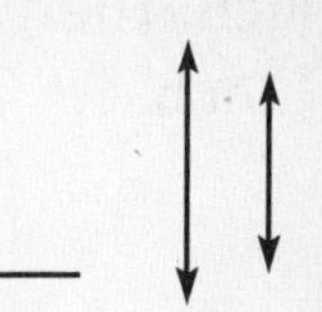

12. ________

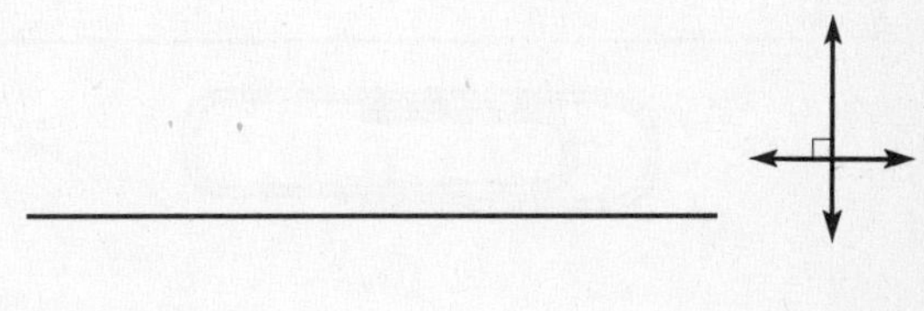

13. ________

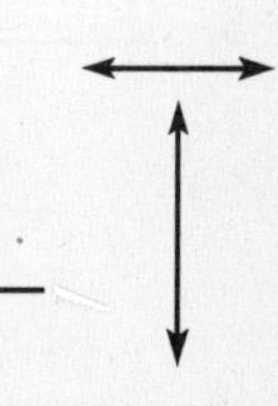

Spiral Review

For 1–3, write the amount as a fraction of a dollar, a decimal, and a money amount.

1. 7 pennies

____

____

2. 5 nickels and 2 pennies

____

____

3. 2 dimes and 1 nickel

____

____

For 4–6, estimate to the nearest inch. Then measure to the nearest $\frac{1}{2}$ inch.

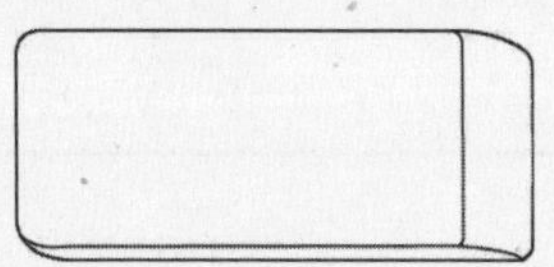

4. ______________________________

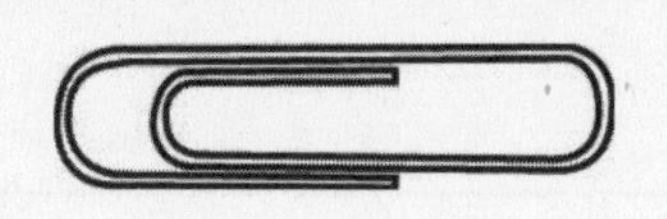

5. ______________________________

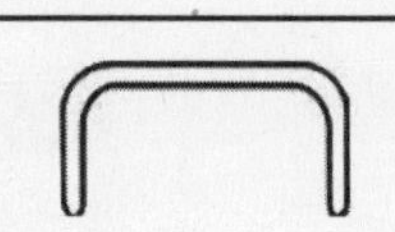

6. ______________________________

For 7–8, make a table to solve.

7. How many combinations of 2 letters can be made from the letters in the word BALL?

_____

B	A	L	L

8. How many combinations of 2 letters can be made from the letters in the word CAT?

_____

For 9–11, tell whether the figure appears to have *no lines of symmetry*, *1 line of symmetry*, or *more than 1 line of symmetry*.

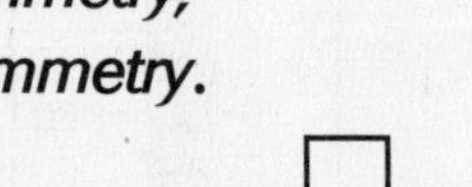

9. ______________________

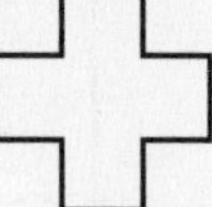

10. ______________________

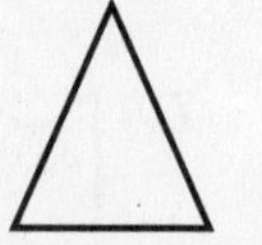

11. ______________________

Name ____________________

Spiral Review

For 1–5, write an equivalent decimal for each. You may use decimal models.

1. 0.5 ________

2. 0.80 ________

3. 0.3 ________

4. $\frac{2}{10}$ ________

5. 0.90 ________

For 6–8, use the calendar below.

July						
Sun	Mon	Tue	Wed	Thu	Fri	Sat
1	2	3	4	5	6	7
8	9	10	11	12	13	14
15	16	17	18	19	20	21
22	23	24	25	26	27	28
29	30	31				

6. How many weeks are between July 8 and July 29?

____

7. Jet celebrated Independence Day on July 4. If today is July 19, how many days has it been since he celebrated?

____

8. Amy's summer reading class starts on July 9. If her class is 14 days long, when will it end?

____

For 9–10, use the double bar graph.

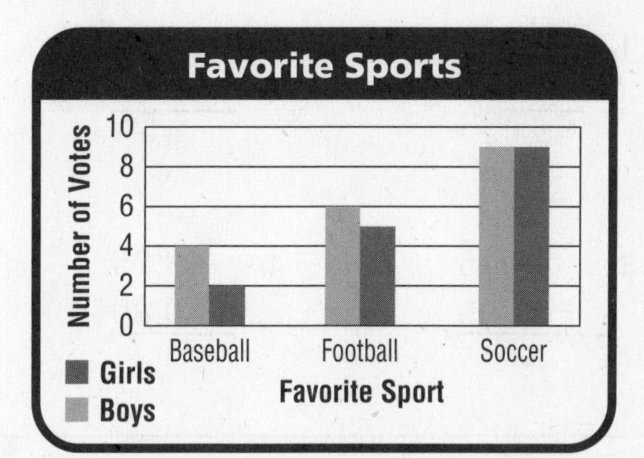

9. Which sport was least popular with boys?

____

10. Which sport was equally popular with the boys and girls?

____

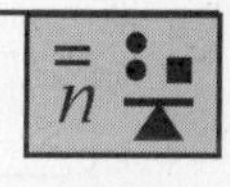

For 11–14, complete. Tell whether you *multiply* or *divide*.

11. 180 in. = ☐ yd ________

12. ☐ pt = 2 gal ________

13. ☐ ft = 18 yd ________

14. ☐ c = 6 qt ________

Spiral Review

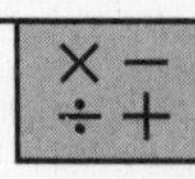

For 1–4, use objects or pictures to find the sum or difference.

1. $5.4 + 1.7$

2. $12.67 - 10.23$

3. $89.45 - 21.96$

4. $\$3.25 + \$0.89$

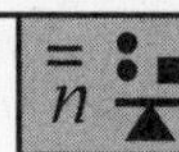

For 5–8, complete. Tell whether you *multiply* or *divide*.

5. ☐ oz = 13 lb ________

6. ☐ lb = 12 T ________

7. ☐ lb = 224 oz ________

8. ☐ T = 32,000 lb ________

For 9–10, use the information below.

Probability Experiment
Tile color: red, blue, green
Pulls: pull 1, pull 2, pull 3

9. Make a model to find all the possible combinations.

10. List all the combinations

____

____

____

____

____

For 11–12, use the table below.

Famous Building	Vertices	Edges	Faces
The Great Pyramid of Giza	5	8	4 triangles, 1 square
Empire State	18	12	6 rectangle

11. What solid figure does the Empire State Building look like?

____

12. Draw a solid figure that looks like the Pyramid of Giza.

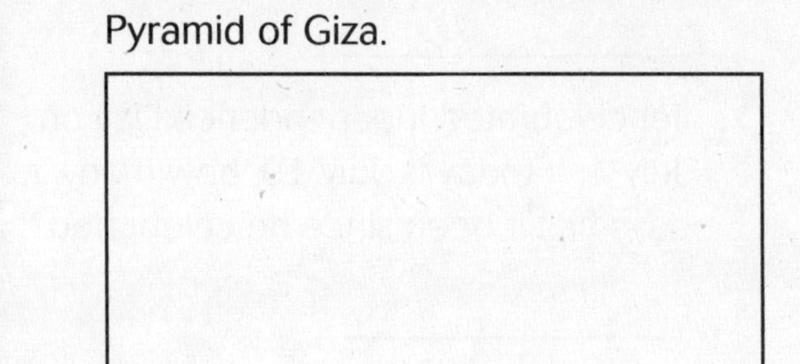

Name ____________________ **Week 25**

Spiral Review

For 1–2, choose a strategy to solve.

1. A rectangular park has a perimeter of 1,200 feet. What is the length of the side *D*? __________

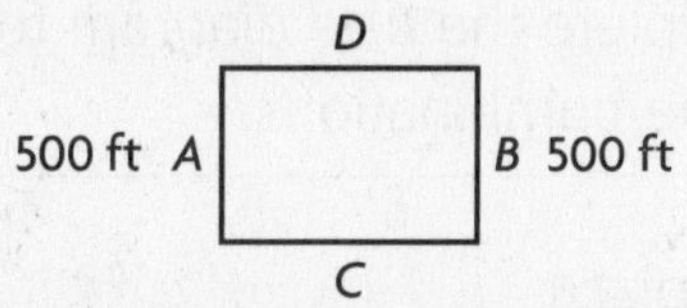

D

500 ft A [rectangle] B 500 ft

C

2. Nikki gets $15 a week for her allowance and completed chores. She spent $7 at the movies. Her friend Arial repaid $2 that she had borrowed from Nikki. Nikki now has $20. How much money did Nikki have to start?

____

For 3–6, choose the unit and tool you can use to measure each.

Tool	Units
Bathroom Scale	oz
Measuring Cup	lb
Meter Stick	kg
Truck Scale	m

3. The weight of a child

____

4. The length of a soccer field

____

5. Water in a glass

____

6. The weight of a car

____

For 7–8, use the table. Make a generalization. Then solve the problem.

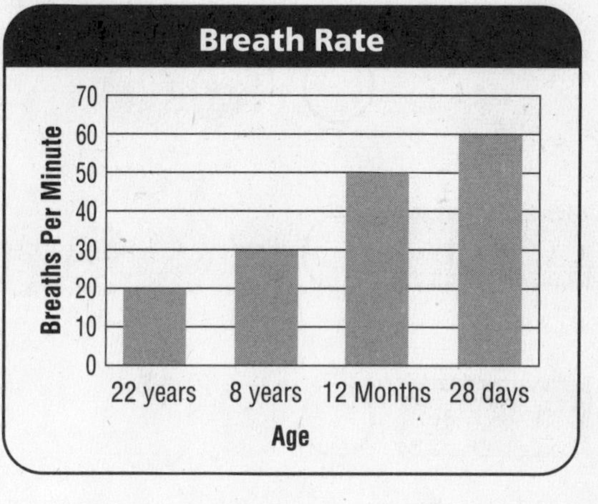

7. Jeff breathes 20 breaths per minute. Gary breathes 50 breaths per minute. Who is older? __________

8. If Jenny breathes 40 breaths per minute. who is the youngest of the three? __________

For 9–12, use the grid. Write the ordered pair for each point.

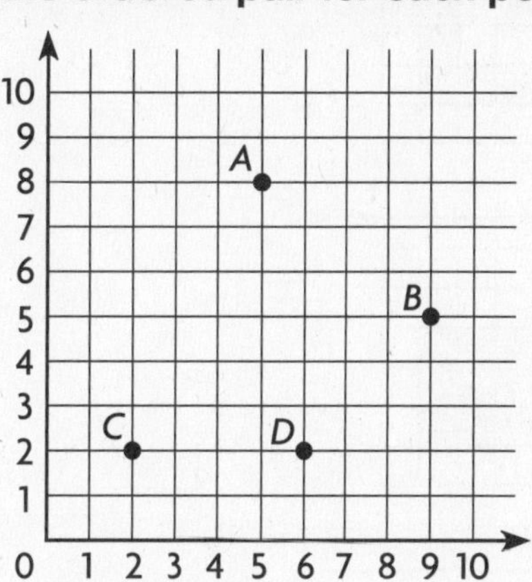

9. Point *A* __________

10. Point *B* __________

11. Point *C* __________

12. Point *D* __________

Spiral Review

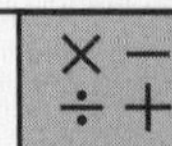

For 1–4, compare. Write <, >, or = for each ◯.

1.    

2.    

3.    

4.    

The school cafeteria has two lunch choices: cheese pizza or chicken nuggets. Each plate comes with the choice of either fruit cup or salad.

8. Complete the tree diagram to show all the combinations.

	fruit cup
cheese pizza	salad
chicken nuggets	fruit cup
	salad

9. List all the different combinations.

____

____

____

____

For 5–7, find the perimeter.

5. 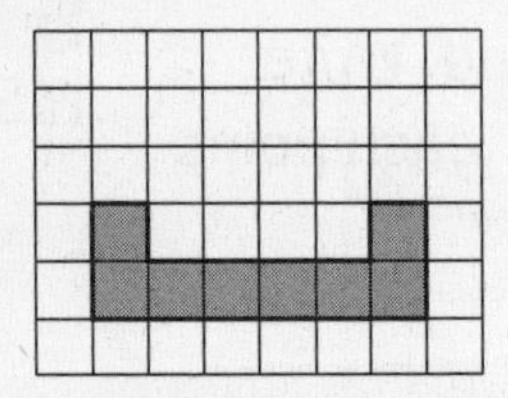

____

6. 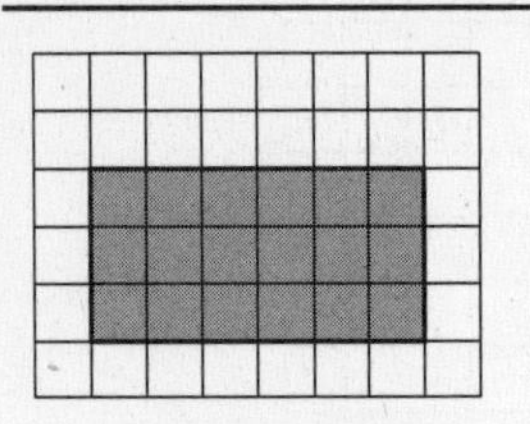

____

7. 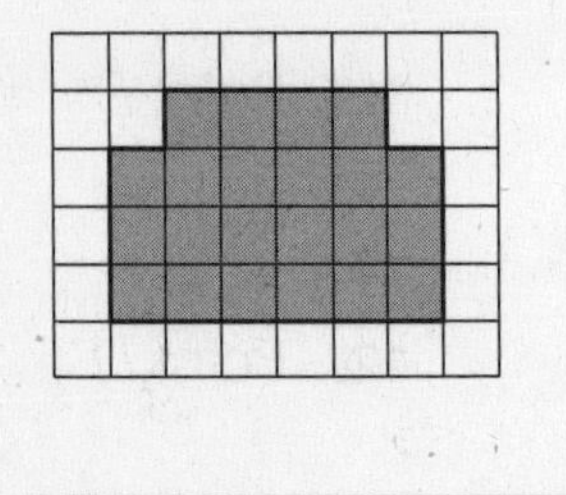

____

For 10–12, Draw each of the following in circle *B* below.

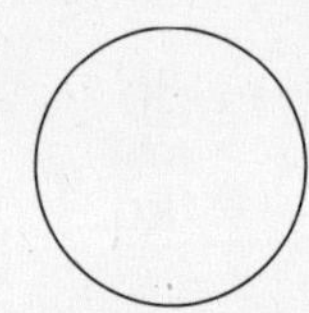

10. Radius: $\overline{AB}$

11. Diameter: $\overline{AC}$

12. Chord: $\overline{DE}$

Name ______________________ **Week 27**

Spiral Review

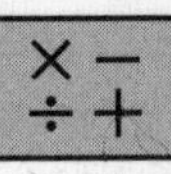

For 1–4, tell whether the fractions are equivalent. Write *yes* or *no*.

1. ____________ $\frac{2}{4}, \frac{1}{2}$

2. ____________ $\frac{3}{8}, \frac{1}{4}$

3. ____________ $\frac{3}{4}, \frac{6}{8}$

4. ____________ $\frac{3}{9}, \frac{1}{3}$

For 5–7, estimate the area of each figure. Each unit stands for 1 sq ft.

5.

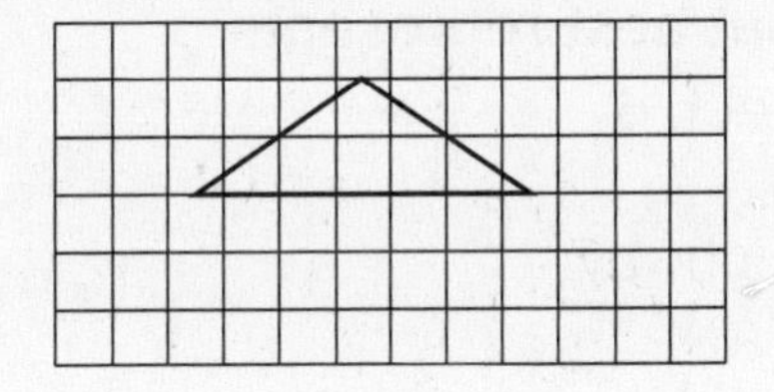

____

6.

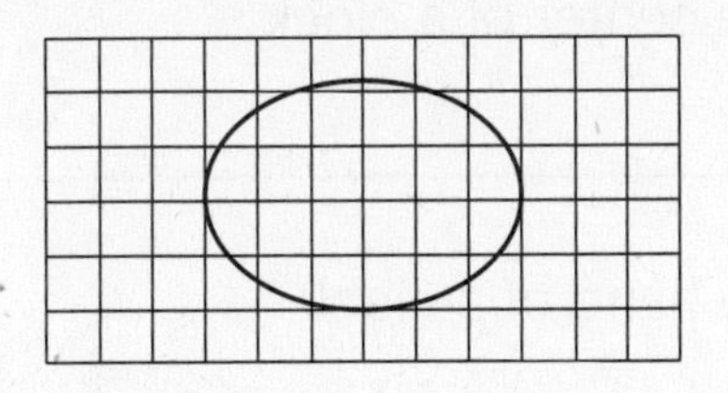

____

7.

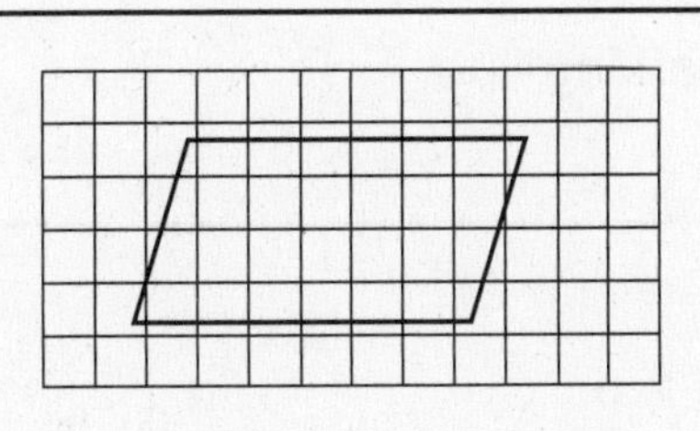

____

8. Audrey is a ski instructor. For her uniform, she is given a red ski jacket and a green jacket. She is also given a red ski cap and a green ski cap. Make an organized list of the possible clothing combinations.

Color of Ski Jackets	Color of Ski Caps

9. According to the organized list you made, how many possible clothing combinations does Audrey have?

____

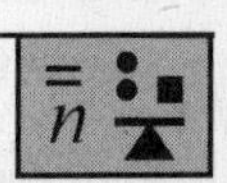

For 10–13, complete. Tell whether you *multiply* or *divide*.

Customary Units of Length
1 foot (ft) = 12 inches (in.)
1 yard (yd) = 3 feet, or 36 inches
1 mile (mi) = 5,280 feet, or 1,760 yards

10. ☐ yd = 324 in.

11. ☐ ft = 108 in.

12. ☐ ft = 10 yd

13. ☐ mi = 10,560 yd

Name ______________________________ **Week 28**

Spiral Review

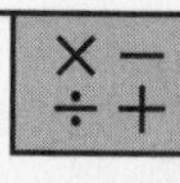

For 1–5, model each fraction to compare. Write <, >, or = for each ◯.

1. $\frac{1}{3}$ ◯ $\frac{2}{3}$
2. $\frac{4}{5}$ ◯ $\frac{2}{5}$
3. $\frac{1}{8}$ ◯ $\frac{2}{9}$
4. $\frac{6}{12}$ ◯ $\frac{1}{2}$
5. $\frac{3}{7}$ ◯ $\frac{8}{9}$

For 6–8, find the area and perimeter of each figure. Then draw another figure that has the same perimeter but a different area.

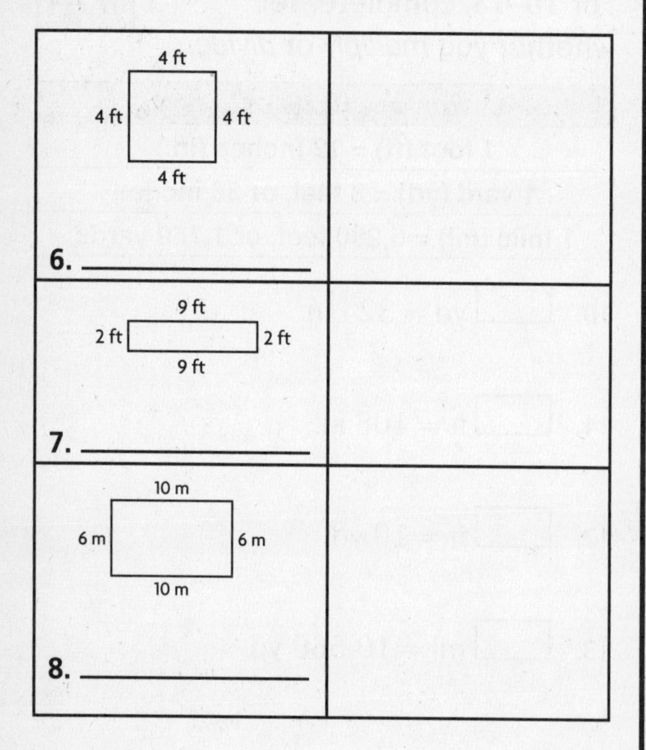

For 9–12, tell whether the data is *numerical* or *categorical* data.

9. color of hair ______________________

10. number of hits in a baseball inning ______________________

11. votes for class president ______________________

12. favorite pets ______________________

For 13–16, name a geometric term that best represents the object.

13. a highway ______________________

14. the center of a clock ______________________

15. the hand of a clock ______________________

16. a parking line ______________________

Name ______________________ **Week 29**

Spiral Review

For 1–3, write a mixed number for each picture.

1.  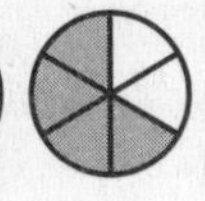

____

2. 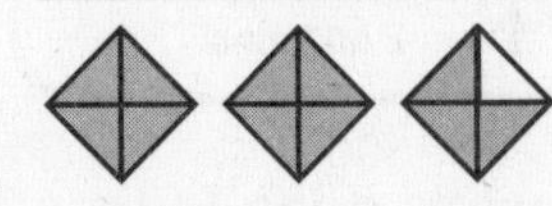

____

3. 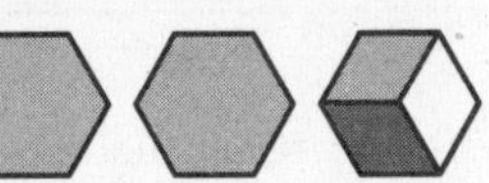

____

For 7–8, complete the Venn diagram below.

A ______ Multples of 4 & 8 ______ B

4, 12 | 16, 8 | 24, 32

7. What labels should you use for section A and section B?

____

____

8. In which section would you put 36?

____

For 4–6, count or multiply to find the volume.

4. 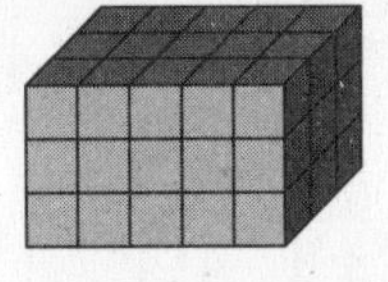

____

5.

____

6. 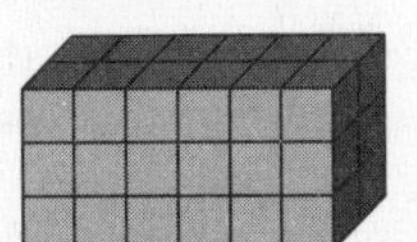

____

For 9–12, complete. Tell whether you *multiply* or *divide*.

Customary Units of Weight

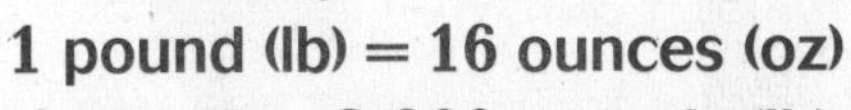

1 pound (lb) = 16 ounces (oz)

1 ton (T) = 2,000 pounds (lb)

1 pint (pt) = 2 cups (c)

1 quart (qt) = 2 pints

1 gallon (gal) = 4 quarts

9. ☐ oz = 10 lb ______________

10. ☐ T = 10,000 lb ______________

11. ☐ qt = 200 c ______________

12. ☐ qt = 20 gal ______________

Spiral Review

For 1–3, compare the mixed numbers. Use <, >, or =.

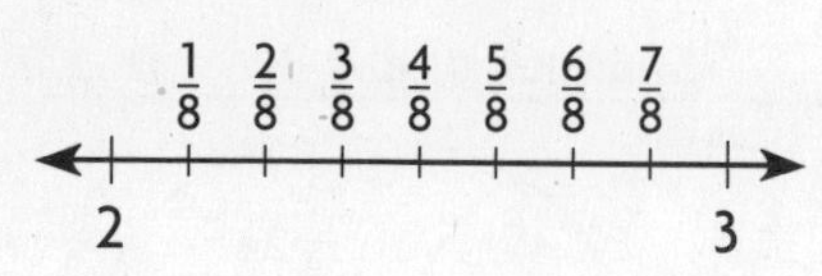

1. $2\frac{2}{8}$ ◯ $2\frac{6}{8}$

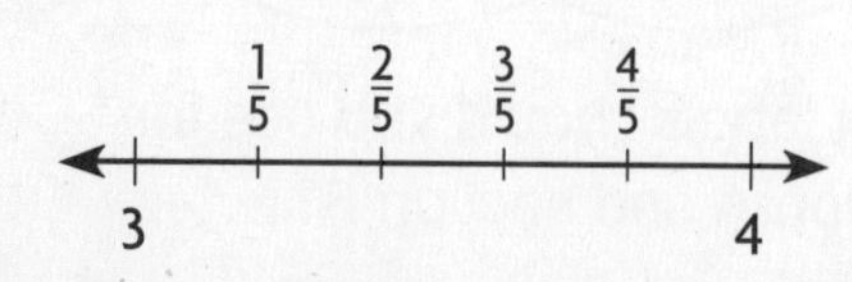

2. $3\frac{2}{5}$ ◯ $3\frac{1}{5}$

$\frac{1}{3}$ $\frac{2}{3}$

4 5

3. $4\frac{1}{3}$ ◯ $4\frac{1}{3}$

For 4–5, compare the volumes of the figures. Write <, >, or = for each ◯.

4.

  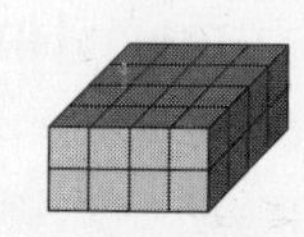

5.

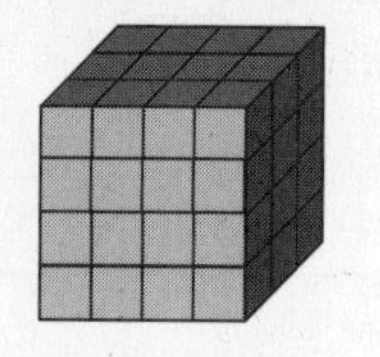 ◯ 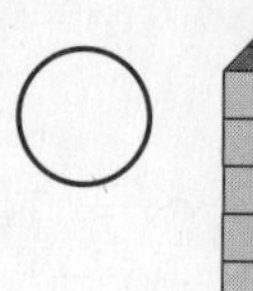

6. Which has greater volume, a prism 3 units long, 3 units wide and 4 units high or a cube with a length, width, and height of 4 units?

____

For 7–9, use the pictograph below.

County Hospital	
Town A	+ + +
Town B	+ + + + +
Town C	+

Key: Each + = 2 hospitals

7. How many hospitals does Town A have?

____

8. Which town has the fewest hospitals?

____

9. How many hospitals are there in all?

____

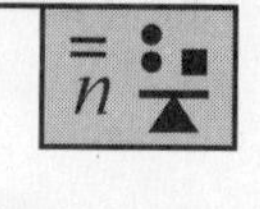

For 10–12, find a rule. Then find the next two numbers in the pattern.

10. 39, 42, 45, 48, 51, 54, ☐, ☐

____

11. 110, 105, 100, 95, 90, ☐, ☐

____

12. 23, 33, 30, 40, 37, 47, 44, ☐, ☐

____

Spiral Review

For 1–4, write each fraction as a decimal. You may use a model or drawing.

1. $\frac{1}{2}$ __________

2. $\frac{90}{100}$ __________

3. $\frac{7}{100}$ __________

4. $\frac{65}{100}$ __________

For 5–8, choose the more reasonable measurement.

5. a nickel: 1 g or 5 kg

____

6. a television: 10 g or 10 kg

____

7. a car: 2,000 g or 2,000 kg

____

8. a soccer ball: 500g or 10 kg

____

For 9–13, choose 5, 10, or 100 as the most reasonable interval for each set of data when drawing a bar graph.

9. 256, 387, 491, 502, 630

____

10. 10, 29, 80, 99, 100

____

11. 10, 129, 180, 199, 310

____

12. 1, 4, 7, 9, 10, 12

____

13. 8, 16, 19, 31, 44

____

For 14–15, classify each angle as *acute*, *right*, or *obtuse*.

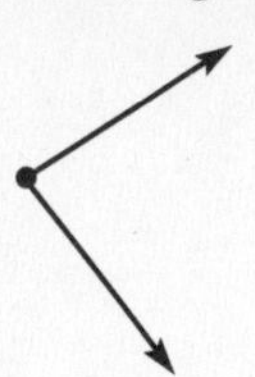

14. __________

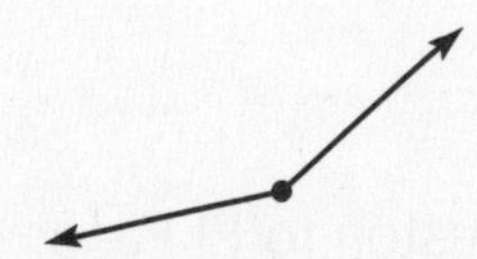

15. __________

Spiral Review

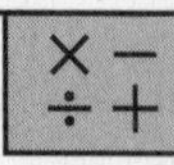

For 1–4, write each fraction as a decimal.

1. $\frac{79}{100}$ ________

2. $\frac{45}{100}$ ________

3. $\frac{3}{10}$ ________

4. $\frac{49}{100}$ ________

For 5–8, choose the more reasonable estimate.

5. a cup of coffee: 250 ml or 25 L

____

6. a shirt button: 1 g or 1 lb

____

7. a school desk: 10 g or 10 kg

____

8. distance from Houston to El Paso: 750 mi or 750 m

____

For 9–11, use the bar graph below.

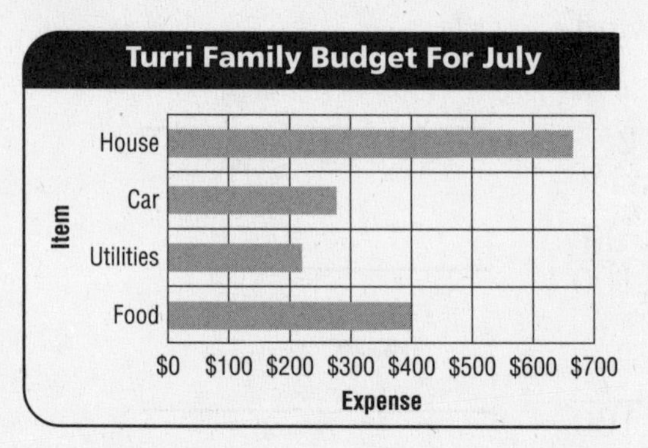

9. How much did the Turris spend on food in July?

____

10. On what did the Turris spend the most money?

____

11. Estimate the Turris' total expenses for the month.

____

For 12–15, use the picture below.

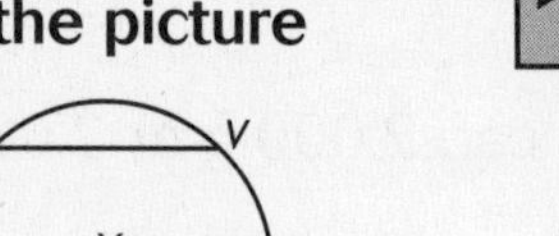

12. Identify $\overline{ZX}$ ________

13. Identify Y ________

14. Identify $\overline{YW}$ ________

15. Identify $\overline{UV}$ ________

Spiral Review

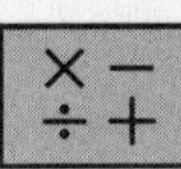

For 1–4, order the decimals from *greatest* to *least*.

1. .7, 1.71, .07

____

2. .05, 5, .5

____

3. .02, .04, .6

____

4. 5.01, 6.99, 6.8

____

For 5–6, find the perimeter of each figure.

5.

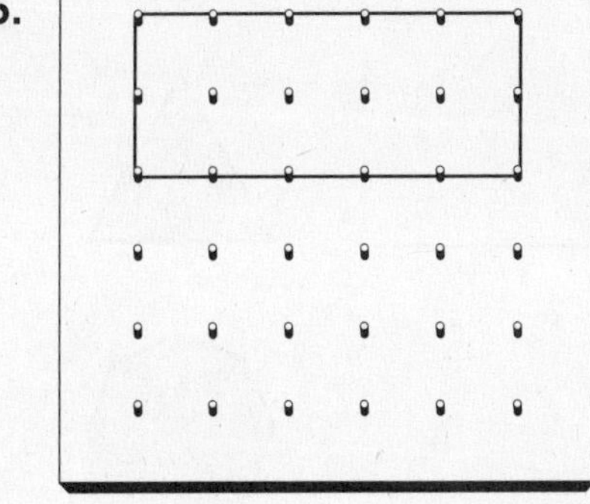

____

6.

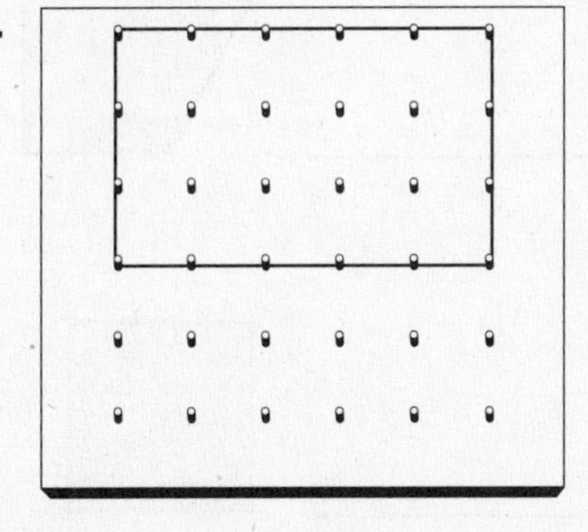

____

For 7–8, use the double-bar graph below.

Number of Votes
35
30
25
20
15
10
5
0
Sports Camp
Drama Camp
Space Camp
Camps
3rd Grade
4th Grade

7. Which camp was least popular with 3rd graders?

____

8. How many more 3rd graders than 4th graders voted for space camp?

____

For 9–11, find the area using the formula $l \times w = a$.

9. ______________________

10 ft
5 ft

10. ______________________

5 in.
5 in.

11. ______________________

9 cm
1 cm

Name ____________________ **Week 34**

Spiral Review

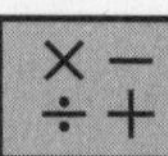

For 1–2, draw conclusions to solve the problem.

1. Jill lives 3.6 miles from school. Gretchen lives 3.1 miles from school. Henry lives 3.7 miles from school. Who lives closest to school?

____

2. Pencils Etc. sells backpacks at 2 for $24.00. Office Box sells the same backpack for $14.00. Which store has the better price?

____

For 3–5, find the elapsed time.

3. Start  Finish 

____

4. Start  Finish 

____

5. Start  Finish 

____

For 6–7, use the word shown below to answer the questions.

HOUSE

6. How many combinations can you make with 3 letters?

____

7. How many combinations can you make with 5 letters?

____

For 8–11, name the polygon. Tell whether it appears *regular* or *not regular*.

8. _______________ 

9. _______________ 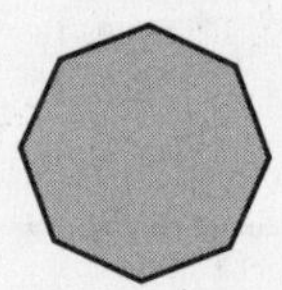

10. _______________ 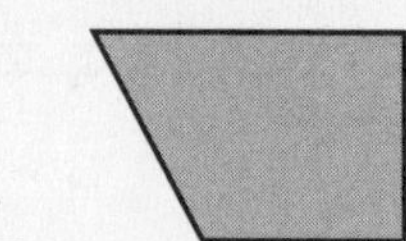

11. _______________ 

Spiral Review

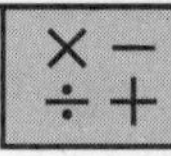

For 1–5, estimate the product. Choose the method.

1. $45 \times 21 =$ ____________
2. $23 \times 11 =$ ____________
3. $30 \times 29 =$ ____________
4. $91 \times 19 =$ ____________
5. $13 \times 13 =$ ____________

For 6–9, use the thermometer to find the temperature in degrees F.

6. 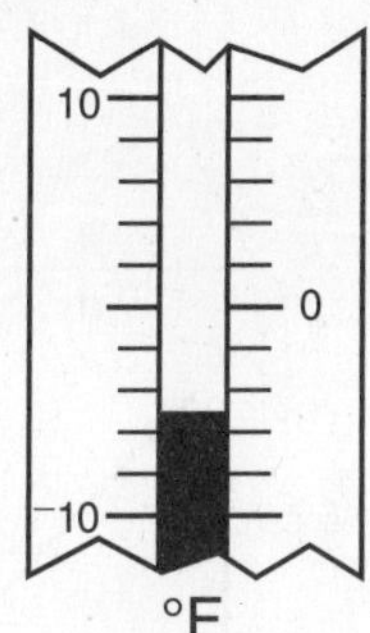

____

7. 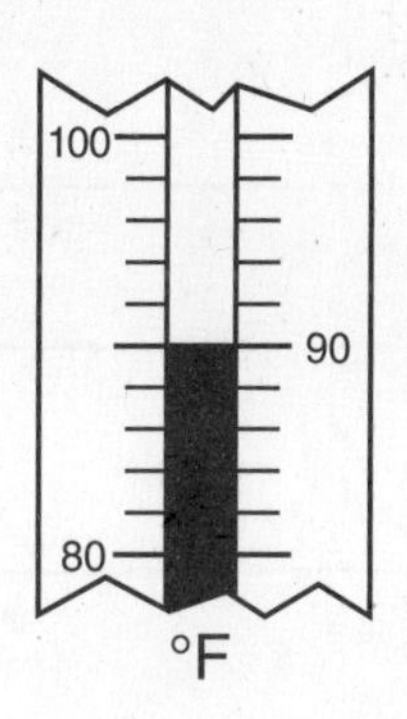

____

8. 9. 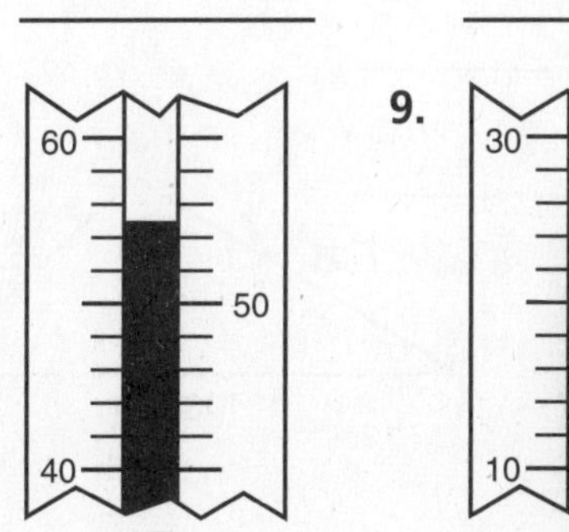

____________ ____________

For 10, make a model to find all the possible combinations.

10. Yoga Classes

time: morning, afternoon

day: Monday, Wednesday, Friday

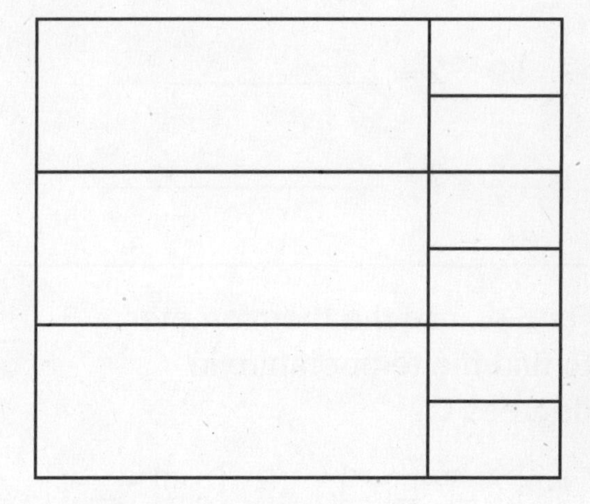

11. How many combinations are possible? ____________

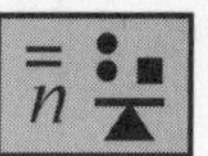

For 12–16, use mental math and patterns to find the product.

12. $9 \times 40 =$ ____________
13. $4 \times 400 =$ ____________
14. $7 \times 3{,}000 =$ ____________
15. $5 \times 6{,}000 =$ ____________
16. $10 \times 100 =$ ____________

Spiral Review

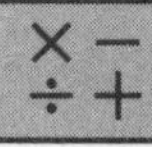

For 1–4, estimate. Then find the product.

1. $31 \times 3 =$ ____________
2. $91 \times 5 =$ ____________
3. $59 \times 2 =$ ____________
4. $22 \times 7 =$ ____________

For 9–12, choose 5, 10, or 100 as the most reasonable interval for each set of data when drawing a bar graph.

9. 41, 73, 31, 88, 24 ________
10. 70, 390, 720, 450, 100 ________
11. 20, 35, 40, 10, 5 ________
12. 250, 300, 100, 200, 500 ________

For 5–8, use the thermometer to find the temperature in degrees C.

5. 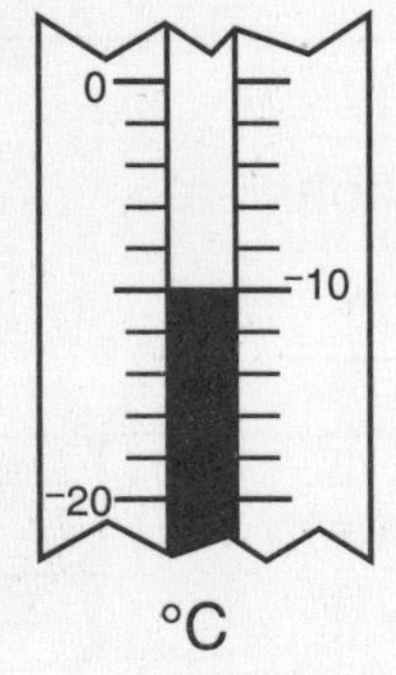

____

6. 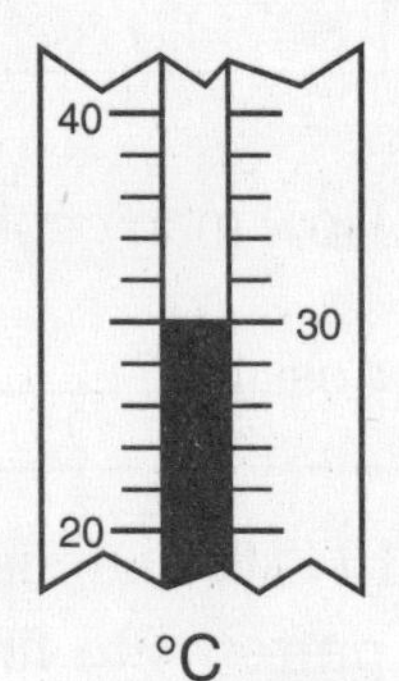

____

7. 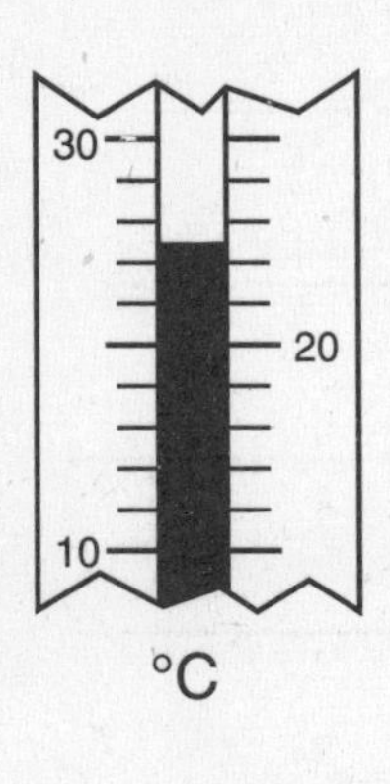

____

8. 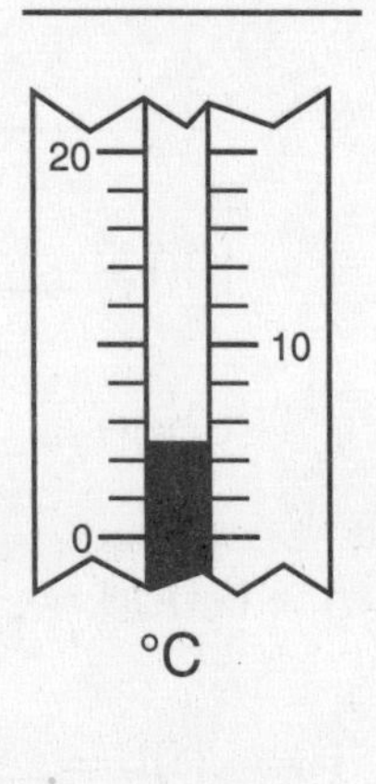

____

For 13–15, classify each triangle. Write *isosceles, scalene*, or *equilateral*. Then *right, acute*, or *obtuse*.

13. ____________

____

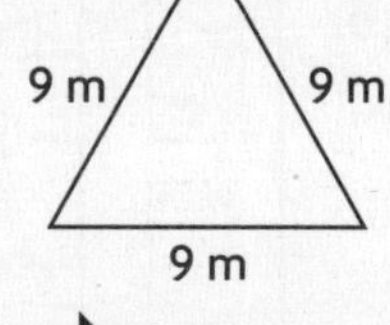

14. ____________

____

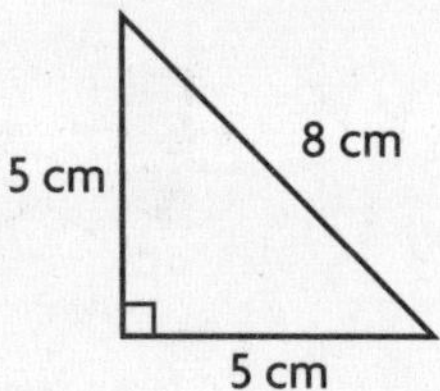

15. ____________

____

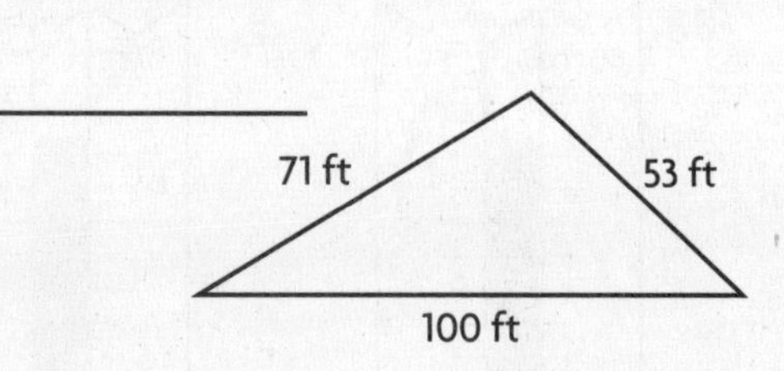